PENGUIN BOOKS

THE HARE AND THE TORTOISE

David P. Barash is a professor of psychology and zoology at the University of Washington. He has contributed to both professional and popular magazines and is the author of eight books, including *The Whisperings Within* and *The Caveman and the Bomb*. He has written and lectured extensively on animal and human social behavior.

DAVID P. BARASH

The HARE *and the* TORTOISE

Culture, Biology, and Human Nature

PENGUIN BOOKS

PENGUIN BOOKS
Viking Penguin Inc., 40 West 23rd Street,
New York, New York 10010, U.S.A.
Penguin Books Ltd, Harmondsworth,
Middlesex, England
Penguin Books Australia Ltd, Ringwood,
Victoria, Australia
Penguin Books Canada Limited, 2801 John Street,
Markham, Ontario, Canada L3R 1B4
Penguin Books (N.Z.) Ltd, 182–190 Wairau Road,
Auckland 10, New Zealand

First published in the United States of America by Viking Penguin Inc. 1986
Published in Penguin Books 1987

LIBRARY OF CONGRESS CATALOGING IN PUBLICATION DATA
Barash, David
The hare and the tortoise.
Bibliography: p.
Includes index.
1. Social evolution. 2. Man—Animal nature.
3. Nature and nurture. 4. Sociobiology. I. Title.
GN360.B37 1987 304.5 86-25392
ISBN 0 14 00.8748 6

Grateful acknowledgment is made for permission to reprint excerpts from the following copyrighted material:

"Sunday Morning" from *The Collected Poems of Wallace Stevens*. Copyright 1923 and renewed 1951 by Wallace Stevens. Reprinted by permission of Alfred A. Knopf, Inc.

"The Bobolink" by Edna St. Vincent Millay from *Collected Poems*, Harper & Row. Copyright 1928, 1955 by Edna St. Vincent Millay and Norma Millay Ellis.

"The Impact on Children and Adolescents of Nuclear Developments" by William Beardslee and John Mack in *Psychological Aspects of Nuclear Developments*, APA Task Force Report #20, Washington, D.C., American Psychiatric Association, copyright 1982. Reprinted with permission.

"Supernatural Songs, XII: Meru" from *The Poems* by W. B. Yeats, edited by Richard J. Finneran. Copyright 1934 by Macmillan Publishing Company, renewed 1962 by Bertha Georgie Yeats. Reprinted with permission of Macmillan Publishing Company and A. P. Watt Ltd.

A speech delivered by Charles Sade at Broome County Community College. Reprinted by permission of Charles M. S. Sade.

Printed in the United States of America by
R. R. Donnelley & Sons Company, Harrisonburg, Virginia
Set in Baskerville

For Nanelle Rose

Contents

It is dangerous to show man too clearly how much he resembles the beast without at the same time showing him his greatness. It is also dangerous to allow him too clear a vision of his greatness without his baseness. It is even more dangerous to leave him in ignorance of both. But it is very profitable to show him both.

—PASCAL

Man is the only animal that laughs and weeps; for he is the only animal that is struck with the difference between what things are, and what they ought to be.

—HENRY HAZLITT

The HARE *and the* TORTOISE

CHAPTER ONE

THE SIAMESE SACK-RACE

It had been the world's first murder. The ape-man exultantly threw his club (actually the leg-bone of a zebra) into the air . . . and as it spun, it turned into an orbiting space station. In this stunning image from *2001: A Space Odyssey*, millions of moviegoers saw the human dilemma in microcosm: we carry the unmistakable signs of our animalness, and yet we do things that have carried us far from the realm of the merely organic. Ape-men all, we are the product of biological evolution—a slow and natural process—and yet we are also enmeshed in our own cultural evolution, which is fast and somehow "unnatural." As the ape-man's club traveled through the air, and ultimately, into outer space, four million years of biological and cultural evolution were collapsed into five seconds.

The journey of the human species dwarfs that of any astronaut, and it is a continuing trip. We are all time travelers, with one foot thrust into the cultural present and the other stuck in our biological past. If we are uncomfortable in this rather awkward posture, no one should be surprised. The human problem is relatively easy to describe but much more difficult to solve: as Pascal recognized so clearly, we are animals, yet we are also much more. The transition from ape to human was dwarfed by the transition from club to missile. Most anthropologists would agree that biologically a prehistoric ape probably wasn't all that far removed from Homo sapiens; it seems a much longer way, on the other hand, from a zebra's leg-bone to a space station. Yet, that remarkable and dizzying transformation occurred almost overnight as evolution reckons these things, and we, we alone, made it happen.

The little hyphen in ape-man is therefore the longest line imaginable, connecting two radically different worlds. Just as Janus the two-faced god was chosen to represent the first month of the Roman calendar because he had one face looking back to the year just past and the other looking ahead to the one forthcoming, our entire species is fundamentally two-faced, one face looking back into our own evolutionary past and the other looking ahead to a rapidly advancing future, all the while delicately balanced in a very transitory present. This book will attempt to make sense of that present, and to suggest a general theory for why it is so confusing, so dangerous, and yet such a hopeful moment in the human adventure.

Our uniqueness as human beings is obvious in nearly every aspect of life, from our stupendous constructions to our most casual conversations. But permeating it all, like the cartoonist's little devil perched on our shoulder and whispering in our ear, is our biological heritage. We have bodies and we have evolved, no less than the ill-fated zebra that our unpleasant ape-man ancestor found so useful. In short, we are biological beings as well as human beings. We bleed, we eat, we defecate, we reproduce, we die. Yet we also dream the most sublime odes, build the most remarkable machines, unlock the atom, imagine eternity and the deity. We are not only *part of* nature, but strangely *outside of* it as well, creatures who in many ways have transcended our organic selves, to think and do things that no other animal ever thinks or does. We can point with pride to our profound accomplishments, and with dismay to the difficulties they pose for us, and for the planet. We are not only Janus-faced, but Janus-souled, riven with a deep-seated dualism that is unique among the earth's creatures. It is both our glory and our curse.

In his *Essay on Man*, Alexander Pope lamented that

> He hangs between, in doubt to act, or rest,
> In doubt to deem himself a God, or beast;
> In doubt his mind or body to prefer;
> Born but to die, and reasoning but to err.

Pope wrote over a hundred years before Darwin, and now we know what the poet did not: We are both God *and* beast, and we do not so much hang between the two as we are stuck in both, simultaneously and irretrievably. Pope concluded that we were a creature of preeminent paradox:

> Created half to rise and half to fall;
> Great lord of all things, yet a prey to all;
> Sole judge of truth, in endless error hurled:
> The glory, jest, and riddle of the world!

Whatever the glory and whatever the jest, Homo sapiens is above all a riddle, the root of which must be proclaimed: the clash between our two fundamental characteristics, culture and biology. This essential dichotomy between the hare and the tortoise, between our galloping culture and our slow-moving biology, is the salient fact of human existence as well as the underlying basis for most of our problems. It is also the basis for this book.

To understand the conflict between culture and biology, we must first review their origins. Our essential bodily characteristics—and presumably, our emotional and mental ones as well—have developed by a gradual process of organic or biological evolution. Although the exact mechanism of this process is still being debated, the essential *fact* of evolution is no longer questioned. Thus, there is "the theory of evolution," and there are "theories of evolution." The various theories of evolution concern the possible mechanisms whereby the evolutionary process is fine-tuned (the role of geological catastrophes, the significance of neutral traits, etc.). But despite the claims of certain Christian fundamentalists, evolution itself is not really a "theory" at all, in the sense of unsupported speculation or a dressed-up hunch; rather, it is as close to fact as science has ever come, analogous to "cell theory," "atomic theory," "gravitational theory," or "the theory of relativity."

As to precisely what role evolution plays in influencing our behavior, there is room for debate. But if evolution is "only a theory," then it is only a theory that the earth is round.

The word "theory" derives from the Greek *theoria*, meaning "viewing or contemplating." A scientific theory is a coherent set of propositions that helps make sense out of facts that would otherwise seem chaotic. It is not a direct line to Truth, but neither is it a wild guess. When it comes to explaining the basic patterns of the living world, evolution has no peers among scientific theories; it does not even have any serious competitors.

Like our biology, our culture has also evolved. The process of *cultural evolution*, however, differs in fundamental ways from the biological evolutionary process that has shaped all life. Our capacity for culture was itself a product of biological evolution, and in this sense, human culture is a direct descendant of our biology. But like an errant child—or Frankenstein's monster—culture developed a momentum of its own and proceeded quite independently of the natural process that originally spawned it. This is because unlike biological evolution, cultural evolution has the capacity to take off on its own, to reproduce, to mutate, and to spread faster and more effectively than any "natural" system. While our biological nature remains shackled by genetics, lumbering along at a tortoise's pace—never faster than one generation at a step, and typically much slower even than that—our culture has been sprinting. In Aesop's fable, the tortoise eventually wins, because the hare is foolish, overconfident, and easily distracted, whereas the tortoise (although slow) is persistent. In the real world, culture and biology differ in speed, but they are equally foolish and equally persistent. And most important, they will both cross the finish line together, because despite their differences they are inextricably tied to each other. We might be tempted to sit back, amused, and watch the entertaining spectacle, sort of a comical, cosmic sack-race featuring two mismatched Siamese twins . . . except that we are part of the show.

The conflict between biology and culture is not a restatement of the ages-old nature/nurture controversy, although it has some similar elements. Biologists, psychologists, and others recognized some time ago that both nature (our genetic and biological heritage) and nurture (our experiences) combine inextricably to produce our behavior. We can no more claim primacy of the one over the other than we can claim that in the structure of a coin, heads are more important than tails. This is equally true for the hare and the tortoise. But although biology and culture are necessarily involved in producing our behavior, they do not necessarily have to get along. When they do, our private lives and our public societies are likely to go smoothly; in such cases, lacking any need for a remedy, most of us are unlikely to notice any need for an explanation.

Fortunately, there can be considerable harmony between our culture and our biology, largely because our biology is so flexible, like a "one size fits all" garment, able to conform to many different shapes and sizes. But not everything fits. Sometimes things just aren't right, and this is distressingly true for the human experience, which has more than its share of warts and rough edges. So whereas all human behavior derives from both biology and culture, both nature and nurture, it does not necessarily follow that biology and culture are always comfortably adjusted to each other. Furthermore, human attention is more likely to focus on conflict than to celebrate harmony, for the same reasons that the daily news reports do not tell us about the things that went well today. Just as we must look to the interaction between nature and nurture for the sources of our behavior, we can look to the conflict between nature and nurture for the sources of our difficulties. A useful rule in murder mysteries—so useful that it became a cliche—is *cherchez la femme* (look for the woman); when Homo sapiens is having trouble, a useful rule—not yet a cliche—would be to look for possible conflict between the hare and the tortoise.

To change the metaphor: two huge continents have drifted apart and now these great tectonic plates, culture and biology, grind together. The results, as we shall see, range from nearly trivial squeaks and wriggles, such as our troublesome sweet tooth or some of our sexual pecadilloes, to the most portentous quakes, such as nuclear war; while in between lie a host of middle sized tremors such as alienation, environmental abuse, and overpopulation. The conflict between culture and biology, the Siamese sack-race between the hare and the tortoise, is an event of paradoxical proportions, ranging from the seismic to the microscopic, from whole societies (indeed, the whole planet and its past, present, and future), to individual people and their likes and dislikes.

Before examining that conflict, in the next two chapters we shall examine the participants, reviewing briefly the anatomy of the tortoise and the hare.

CHAPTER TWO

ANATOMY OF THE TORTOISE

We live in an old chaos of the sun,
Or old dependency of day and night,
Or island solitude, unsponsored, free,
Of that wide water, inescapable.
—WALLACE STEVENS,
"Sunday Morning"

Biological evolution is the ultimate natural process. It is the great flow of events that created all living things and also the thread of continuity that links them together even now. For us, as for all living things, it is an old dependency. And it is inescapable. Julian Huxley once pointed out that the human species is evolution becoming aware of itself. However, most Homo sapiens didn't partake readily of this awareness, in part because the subtlety and slow pace of biological evolution make it difficult to identify, in part because of discomfort over a scientific truth that ran counter to established religious doctrine, and in part out of reluctance to admit a unity of human beings with "bestial" nature. "Descended from monkeys?" exclaimed the wife of the Bishop of Worcester in the mid–nineteenth century, "My dear, let us hope that it isn't true! But if it is, let us hope that it doesn't become widely known!"

Such reactions have a long history, and they often came gift-wrapped in elaborate intellectual paraphernalia as well. For example, after Copernicus had upset the comfortable, ego-pleasing, and universally accepted notion that the earth was the center of the universe, the Danish astronomer Tycho Brahe did his best to see to it that a more congenial view became widely known, even if it wasn't true. Brahe proposed that the five identified planets rotated about the sun, as Copernicus had demonstrated, but that the whole business, in turn, went around the earth! Brahe was not consciously seeking to deceive his contemporaries; he sought a view of the universe that was more in accord with what he wanted to be true.

Such a proposal was not, strictly speaking, a compromise, analogous to "splitting the difference" between contending parties. Rather, the Brahean solution was an ingenious but ultimately unsuccessful example of seeking to accommodate to new and awkward facts by accepting those that are undeniable—so long as they are marginal to the most deeply held beliefs—while clinging tenaciously to what is dearest to us, even if it may be incorrect.

Human beings periodically seek Brahean solutions. We try to adapt to new data, while nonetheless keeping our fundamental orientation intact: we grant the laws of physics, while leaving room for free will; we grant the finite limits to the earth's resources, while continuing to exploit them; we grant that nuclear weapons are unusable, while continuing to build more and more. Our view of human nature has also been rather Brahean: most people today accept evolution, but still maintain a conception of themselves as separate, unique, and distinct, God's special representatives on earth, if not in the universe. Like the inventive Tycho Brahe, we grudgingly accept certain unavoidable facts—such as our biological connectedness to other living things, notably reflected in paleontology, anatomy, embryology, and physiology—while strenuously resisting the heresy that in our behavior, too, we may be similarly connected to the other inhabitants of our planet.

Perhaps we should relax, since in fact our uniqueness is comfortably assured, even though on first glance it may seem otherwise. Our planet, admittedly, is insignificant compared to the rest of the universe. Not only do we circle the sun rather than vice versa, but even old Sol is rather small, out of the way, and second-rate by astronomical measures. And even here, on the third planet from the sun, not only are we animals, but as one species among millions, Homo sapiens is about as insignificant as the earth is among the heavens. However, when we examine the realm of life itself, the telescope is inverted: the universe, cold and perhaps utterly lifeless, is no longer vast and imposing, but instead, trivial compared to the marvelously living, breathing creatures that grace the earth. And in a sense, all these creatures pale by contrast with the human species, whose consciousness and culture make him and her something special indeed, something really new under this sun and perhaps all suns.

Lewis Mumford has written that "without man's cumulative capacity to give symbolic form to experience, to reflect upon it and refashion it and project it, the physical universe would be as empty of meaning as a handless clock; its ticking would tell nothing. The mindfulness of man makes the difference."

Our "mindfulness" is also a product of evolution, undoubtedly the greatest triumph of organization and complexity that a fundamentally unmindful, unsponsored process has yet created. "If this property of complexity could somehow be transformed into visible brightness," writes molecular biologist John Rader Platt, in *The Step to Man*,

> the biological world would become a walking field of light compared to the physical world . . . an earthworm would be a beacon . . . human beings would stand out like blazing suns of complexity, flashing bursts of meaning to each other through the dull night of the physical world between.

Perhaps Brahean solutions to our self-perception and especially our relation to the physical and biological world are less needed today than ever before, as we become increasingly aware of the universe as a handless clock, and, by contrast, of the mind-boggling complexity of life in general and of the human mind in particular. Such awareness is due at least partly to the welcome fact that our perception of "nature" is more sophisticated than ever before. The modern (and presumably, more accurate) view of nature is that it is neither the mysterious "élan vital" of an Henri Bergson, nor the rigid, clockwork automaticity of a Descartes, or a world of Newtonian billiard balls. Rather, in a curious merger of wisdom East and West—the world's great mystical traditions as well as modern physics and ecology—science today increasingly sees nature as a dynamic, ever-changing, permeable state of exchange and equilibrium; that is, nature is increasingly recognized to be neither a fixed, linear sequence of set levers and pulleys, nor a mystical bowl of amoeboid mush, but rather, a *process*. Joining the melee, recognizing ourselves as part of the process, we are neither diminished nor expanded, just described. So the need for a Brahean solution is thereby reduced, and we emerge freer than ever before to see ourselves not as others see us or as we might wish to see ourselves, but as, perhaps, we really are.

DARWIN did not "discover" evolution. It had been described and speculated upon by numerous authors before him, including his own grandfather. Rather, his great contribution was to identify a mechanism whereby evolution can plausibly be seen to have occurred: natural selection. Like so many great intellectual discoveries, natural selection is a logical necessity, given certain basic facts. Darwin's particular genius lay in recognizing the significance of commonplace, commonsense observations, just as it took a special mind for Isaac Newton, generations earlier, to divorce himself from the commonplace observation that things fall, and to view gravity as something worth examining. "Of course," commented Thomas Huxley upon reading *The Origin of Species*, "how stupid of me not to have thought of that!"

Natural selection—and hence, evolution—is simply a logical consequence of the way the biological world is constructed. In fact, it probably cannot be avoided. To maintain a constant population, individuals of any sexually reproducing species (with one male and one female as parents) must simply produce two surviving offspring, thereby replacing themselves. Fewer than this will eventually cause extinction; more will cause an increase in numbers, and the increase would itself continue to increase. Certain species such as the lynx and snowshoe hare in northern Canada or the lemming of Scandinavia experience cyclic increases in numbers followed by spectacular decreases. But by and large, the natural populations of most animals are balanced, showing only minor fluctuations in numbers. This indicates that in most cases, parents simply replace themselves, although most living things have the capacity to produce many more offspring than this. Thus, a female cod may produce a million eggs at a single spawning; if all these were to survive and then each of the million reproduce in turn, the oceans would soon be too small to hold all the codfish alone.

Even a slower breeder such as the robin, producing say four eggs at a sitting, can generate sixteen offspring in just four years, and if each of these offspring also reproduces comparably, our original pair will have been responsible for 104 direct descendants in this same four years. Or, the uncontrolled reproduction of houseflies would soon theoretically produce a ball of flies larger than the earth. But such numbers are hypothetical. These statements are mathematically correct; the fact that such dire predictions do not come true is testimony to the high mortality of living things in nature, and to the fact that reproductive potential is only rarely realized. All the potential offspring of most species do not survive. In fact, as we have seen, just two young of every species will eventually live to replace their parents, if the population is to remain constant, as most populations do. This means that 999,998 aspiring cod must perish annually along with 102 potential robins every four years, for two of each species that survive.

What separates the winners from the losers? Those that are better adapted to succeed will do so. In fact, this is what we mean by "better adapted." They will be *selected* by *nature* as the chosen representatives to propagate the next generation: in short, natural selection. As Darwin recognized, natural selection thus follows inevitably from the capacity of living things to reproduce in great numbers, combined with the fact that very few of the potential newcomers are actually successful.

This process could not produce change, however, unless the offspring from which nature selects differ genetically from each other. Thus, even a continual winnowing cannot achieve anything new if the components are all fundamentally the same. Only in the case of identical twins are two individuals of a sexual species genetically identical. With these rare exceptions, each individual is genetically distinct, each with its own particular makeup, its own particular contribution to make, and its own probability of failing. Its ultimate success depends on whether it is "fit"—that is, well adapted to survive and reproduce—and this fitness will be influenced by the unique gene combination each particular individual is carrying. Successful gene combinations are thereby selected for, eventually replacing their less successful relatives.

When Darwin first described natural selection, it was almost universally *mis*understood: it was thought to operate by violence and death. Evolutionary change was seen as the necessary outcome of Tennyson's "nature red in tooth and claw." "Social Darwinism" became a convenient credo of late-nineteenth-century laissez-faire capitalism. Under the excuse that "survival of the fittest" provided biological sanction for the most abhorrent social practices, domination of the weak by the strong was proclaimed as natural and right, a law of nature and hence, a dictate from God. Actually, the outcome of aggressive fights—waged either by tooth or stock option—is almost irrelevant to natural selection. As Darwin himself seemed to recognize, individuals are not selected for, *genes* are. To some extent then, the victor in a fight to the death might enjoy a selective advantage over an opponent, but only insofar as he or she would be more likely to leave successful offspring as a result. And even more important, advocates of social Darwinism fell victim to what David Hume had earlier called the "naturalistic fallacy": what is, ought to be. The social Darwinists misread not only science, but ethics as well.

Natural selection is not a prescription of how the world ought to be; it is a *description* of how it works and of the major way in which evolutionary change comes about. Natural selection is simply differential reproduction, primarily of individuals and their genes. Within any species, individuals leaving a larger number of offspring are being selected for—or rather, their genes are being selected, since they will enjoy a greater representation in succeeding generations. Those with fewer offspring, and fewer genetic copies projected into the future, are selected against. It's that simple. Evolution by natural selection favors any genetically influenced characteristic that increases the chances of its carriers leaving more offspring and other genetic relatives. These characteristics may include success in personal conflicts, but more likely, ability to prosper in their respective environments. Thus, the ability to find food, avoid enemies, attract a mate, obtain shelter, withstand climatic stress, and cooperate with others may all enjoy natural selection's favor, because possessors of these traits would be more likely to leave progeny that would carry similar tendencies and that would be more successful in turn, thereby increasing the frequency of these genes in the population. Natural selection thus proceeds inexorably from the simple fact that organisms are capable of great overproduction, while population size generally remains relatively constant; then, given genetic differences among individuals, selection will favor those traits that contribute most to reproduction in the particular environment in question.

The sources of these genetic differences, however, have only recently been identified. The fundamental building block of genetic variety is now known to be mutation, basically a stenographic error in the cell's copying machinery. The genetic composition of every living thing is coded by the specific arrangement of atoms in complex organic molecules, the nucleic acids (so named because they are most abundant in the nucleus). These nucleic acids—DNA in most animals—are unique to each species and must be copied precisely every time a cell divides, to insure that the daughter cells will be like the parent and ultimately to guarantee that watermelons produce watermelons while people produce people.

Usually this copying is remarkably exact, but occasionally an error makes the copied DNA very slightly different from the original, like a typed manuscript in which the wrong key was struck. In most cases, a mutation reduces the fitness of the living thing that carries it, just as an error generally reduces the quality of a typed work. Or imagine a delicate, carefully balanced machine: it is very unlikely that a random change will improve its operation. But occasionally, the error may actually improve the manuscript, by suggesting a better word than the original. Very rarely a mutation may actually improve performance and therefore be selected for. Perhaps one mutation in a thousand will be beneficial, while the mutations themselves occur about one in a *million*; nature is a very competent typist. At this point, however, the typewriter analogy breaks down, because mutations may involve not only the substitution of the wrong letter in a complex manuscript, but in fact the creation of an entirely new letter, thus potentially expanding the genetic repertoire, the "alphabet" of the species.

Mutations, like most other errors, are random in that specific instances cannot be predicted in advance. Like other errors, however, they are not entirely unpredictable either. Certain genes are more likely to mutate than others and although the exact time of any given mutation cannot be predicted, statistical estimates can be made. Thus we may say that gene "X" will mutate, on the average, once in one million copies, while gene "Y" will mutate once in ten million, just as insurance analysts will estimate different accident frequencies for different industrial operations. Furthermore, and once more like other accidents, mutation rates are influenced by external factors, in this case, certain chemicals (mustard gas for example), excessive heat, radiation (ultraviolet light, X-rays, or nuclear radiation), or even the presence of special genes that influence the mutation rate of other genes.

During their evolutionary history, most living things did not experience high-energy radiation; not surprisingly, therefore, they are vulnerable to its effects. This susceptibility of the genetic copying process to high-energy radiation is the basis for biologists' concern about excessive medical X-rays or radioactive nuclear fallout. In the absence of these artificially induced goads to higher mutation, enough genetic copying errors occur naturally, presumably in response to the normal imperfection of the material world or to natural factors in the environment, to provide evolution with a constant supply of genetic variation from which natural selection can select. It is sobering for human perfectionists to consider that all biological advance is ultimately based on error, nature's very essential fallibility.

Evolution by natural selection is a painfully slow process. Modern-day ferns, for example, are basically unchanged over hundreds of millions of years. The same goes for horseshoe crabs, turtles, and crocodiles. Even those species that have evolved "rapidly," such as horses, elephants, or human beings, took many hundreds of thousands and more often millions of years to progress from horselike, elephantlike, and humanlike ancestors to the forms we recognize today. Compared to the rate of cultural change, everything is a "living fossil."

Evolution would be even slower yet if it had to rely on mutation alone, since mutations are so very rare. Most living things, human beings included, long ago stumbled upon a wonderful device for producing a great variety of genetic combinations in every generation: sex. With sexual reproduction, genes from each parent are combined and recombined, organized and reorganized every generation, producing a unique genetic makeup for every offspring. This explains the seeming contradiction that whereas like always begets like—human beings never give birth to giraffes—children are never truly identical to their parents.

This, then, is the crucial biological consequence of sex: not reproduction per se (after all, many living things reproduce asexually), but rather, the creation of a vast store of genetic variation upon which natural selection can operate. Sexual reproduction shuffles and reshuffles the cards before every child is dealt his or her genetic "hand." In this way, mutations first appearing thousands of generations ago can be retested in different contexts until either a winning combination is reached or the hand is declared a failure. If successful, a combination will be favored by natural selection, selected *for*, and evolution will proceed by increasing its representation in the population; carriers of that combination will leave more offspring. Holders of losing hands, in turn, will leave fewer offspring. They will be selected *against*, and their combinations will eventually retire from the game, unable to compete with the winners. If everyone in the species is dealt a bad hand, or if the rules are changed too quickly, the species goes extinct.

Genetic variation increases the range of diversity shown for any characteristic. Such diversity can be seen, for example, in the wide range of human height from short Eskimo to tall Watusi, or even the varying statures of children from the same parents. We vary in all traits—from shoe size to intelligence—and this variability is in part an expression of the genetic variation that distinguishes each individual.

But why should such variation exist at all? With natural selection constantly pruning out the less successful variants, one might expect each species to eventually consist only of "super(wo)men," perfectly adapted to their environment. To some extent this is prevented by the very fact of error itself: mistakes do happen, ultimately beneficial or not. But beyond this, optimum characteristics for a species are always relative to its environment; so characteristics advantageous now may be of no value or even become liabilities if the environment changes . . . and eventually, it always does. Thus, keen eyesight might seem universally advantageous, but not to an animal inhabiting caves where it is perpetually dark. For such animals, eyes are not only useless but an actual hindrance since they are delicate and prone to injury and infection. Those fish and salamanders whose ancestors took up cave dwelling as a way of life have profited from the fact that they retained the genetic variety to produce eyeless forms when necessary.

Since selection tends to eliminate the unfit, and ultimately even the *less* fit, a greater diversity of types is present in each species before selection operates than after. Each species therefore experiences conflicting forces: mutation and sexual recombination acting to increase the range of variability, and natural selection generally tending to narrow this range. The pruning action of selection alone, without the disruptive effect of mutation and recombination, would tend to produce a somewhat uniform population, well adapted to a particular environment but vulnerable to change. Mutation and recombination, without the narrowing effects of selection, would produce an array of freaks, poorly equipped to be superior anywhere but perhaps with some representatives capable of surviving almost anywhere. Each species therefore arrives at its particular evolutionary strategy, combining varying amounts of short-term success with long-term insurance. There are numerous complex techniques for preserving hidden variability, keeping it from the probing eye of natural selection. For example, genes are normally found in pairs, of which one may be "dominant," overshadowing the influence of the other. In this way, a deleterious "recessive" mutation may be providentially retained in the population, only to prove advantageous—or alternatively, to be selected against—in the future. Among human beings, for instance, the tendency to produce blood that clots normally is controlled by a dominant gene; hemophilia is produced by its alternative, recessive form. Two seemingly normal people can therefore produce a hemophilic child if they are both carrying a recessive gene, masked by a dominant, and if the unlucky child gets a dose of the recessive gene from each parent.

If we reproduced asexually, those of us with normally clotting blood would be assured of producing offspring whose blood also clotted normally, since our offspring would be identical to ourselves. But by turning our backs on sexuality, we would also have foregone the opportunity of combining our genes with those of another person, an adventure whose outcome can be offspring that are new and guaranteed to be different, possibly less "fit" than ourselves, but possibly more so. Most of the higher animals, our own species included, have taken the plunge. We have opted for sexuality, with its consequent genetic variety and promise of evolutionary flexibility. By contrast, certain living things have been more conservative, mortgaging the future for a high degree of present-day fitness. The common yellow dandelion, for example, has foresaken sexual reproduction in return for immediate success. But when the environment changes, dandelions may well lack the genetic reserves needed to adapt to a new situation. In this regard, sexual abstention is in fact a profligacy of sorts, one for which the indulgent must eventually pay the price. Other living things, such as daphnia (water fleas) or aphids (plant lice), attempt an intermediate strategy. They rely largely on asexual reproduction, but they interject an occasional bout of sexuality each year, to reshuffle the cards just in case.

Ultimately, even a very sexy species may find itself unable to adapt to a changing environment and thus go extinct. In such situations the species is "overspecialized," analogous to the overspecialized worker whose technical training leads to unemployment. Overspecialization, in fact, is probably the most common cause of extinction. For example, saber-toothed "tigers" weren't really tigers but large, heavy-bodied scavengers whose enormous canine teeth probably served to pierce the thick hide of dead mastodons, mammoths, and titanotheres on which they fed. By the time their prey disappeared, these remarkable cats, saber teeth and all, apparently had gone too far down the road of specialization to adopt a new way of life. They were trapped, organisms without an environment, and are now extinct. Similarly, the sleek and graceful Everglades kite (a bird) feeds only on one species of snail, found in southern Florida. As Everglades National Park suffers increasing drought and other water problems compounded by human interference, the snail will probably disappear, and with it will go the Everglades kite.

By contrast, Homo sapiens is a relatively unspecialized animal, capable of exploiting a tremendous variety of life styles, "niches" to the professional ecologist. As we engineer rapid and far-reaching changes in the world's environments, we can look ahead to the likely disappearance of the world's specialized organisms—koala bear, condor, giant panda, and tiger—and continued success of the unspecialized: the raccoon, starling, rat, housefly, and cockroach. Keep in mind, however, that our own generalized biology combines with a highly specialized reliance on an extraordinary characteristic: culture. Some specializations, like the saber-tooth's canines, boxed their possessors into a cul-de-sac. They were tickets to doom. Others, like the peculiar reptiles that developed their forelimbs into wings and their scales into feathers, eventually giving rise to the wonderfully diverse order of birds, were the start of something big: a whole line of evolutionary success. Our cultural specialization has clearly taken flight; it remains to be seen whether we shall soar smoothly and safely like the descendants of *Archaeopteryx* or follow instead the precedent of Icarus, child of Daedalus, who flew too near the sun on his manmade wings, melted their wax, and crashed in ruin.

It is important to remember that evolution proceeds by the gradual accumulation, substitution, and exchange of favorable genes within individuals. Note that individuals do not evolve; species do. An individual is born with a particular genetic constitution and cannot change it, any more than the Everglades kite can shed its single-minded passion for snails. Only by reproduction are new genetic possibilities created, from which natural selection will choose some to bask in the future. The rate of evolutionary change will therefore depend on several different factors, notably the diversity present within a given population, the extent to which this diversity is a reflection of underlying genetic diversity, and the "selection pressure," the extent to which individuals and genes of one sort are reproductively favored over others. Even under optimum conditions for rapid change, evolution must await the passing of many, many generations as environmental pressures gradually knead and shape the gene pool of a species.

How evolution works is one thing; what it has done is another. No one really knows how life first appeared on earth. Because the event is shrouded in about six billion years of time past, it is entirely possible that we shall never know. Conflicting testimony at any legal proceeding will reveal that the more distant an event in time, the greater the disagreement about its details. As one of the most ancient events, the origin of life is no exception. However, we can skip the complex chemical formulations and describe one hypothetical sequence of events that has been accepted by a large number of experts. The early atmosphere probably consisted largely of hydrogen, nitrogen, water vapor, methane, and ammonia. By placing these substances in a constantly circulating system, applying occasional electric sparks (simulating ancient lightning) and ultraviolet radiation (simulating the sun), a great variety of complex organic molecules can be produced in the laboratory. Among these organic molecules are several amino acids, which are the basic components of proteins and even the precursors of nucleic acids themselves. This experiment may re-create the basic steps that actually took place so long ago; at least, it shows that inorganic compounds plus energy can produce organic compounds. As more and more complex molecules were formed, they would have interacted with each other, until a virtual "soup" of organic chemicals was produced. Given the rich consommé, it would only be necessary for one array to develop the ability to reproduce itself (as the DNA molecule does now in each of us) for life to have appeared.

The primitive life forms that were first able to reproduce were presumably surrounded by other organic molecules, generated by similar forces, but not quite at the "living" stage. This rich broth would have provided an abundant food source for any of our molecular ancestors that could take advantage of the energy accumulated in their structure. In a sense, therefore, the first living things were animallike in that they obtained their nourishment by "eating" other, dead things.

As the early animals began to slurp up their organic soup, natural selection presumably began to favor genetic variants that could produce their own food, using only the bare resources of carbon dioxide, water, and the energy of light: the first plants. As they diversified, plants not only provided an enhanced food supply for the increasingly hungry animals but by the process of photosynthesis (chemically joining carbon dioxide and water to produce glucose, a simple sugar) they added a new and important component to the atmosphere: oxygen. The effect of oxygen was twofold. By absorbing the sun's abundant ultraviolet radiation, oxygen was converted into ozone, which, present in the upper atmosphere today, screens out a high percentage of the ultraviolet radiation reaching the earth. Without this ozone layer, high levels of ultraviolet light would produce an increased frequency of skin cancer, blindness, and genetic mutations.

The second major impact of atmospheric oxygen was to provide a basic chemical environment in which life could be much more lively. Without oxygen, organisms would have to utilize a rather inefficient energy pathway, one that is employed today only by certain primitive organisms—such as yeast, which conveniently produce alcohol as a by-product—and by higher animals only for short periods at a time. An athlete's muscles can produce energy without oxygen during intense stress; accumulation of the end product of this process, lactic acid, is felt as a muscle cramp. When exercise is over, rapid breathing pays off the "oxygen debt" built up earlier. Limited to an oxygenless environment, animals were restricted to an uninspiring, sluggish existence. Freed from these metabolic strictures, however, they were ready for a riot of activity: individual and evolutionary, and given enough time, cultural as well as biological.

Early life was probably limited to the primitive seas. The first vertebrates evolved several hundred million years after the earliest invertebrates and were initially dwarfed by their highly evolved but boneless cousins. Trilobites (resembling present-day horseshoe crabs) and occasional giant eurypterids (enormous ocean-going scorpions) dominated the early oceans while small, inconspicuous vertebrate fish probably skulked in the shadows. These early fish even lacked movable jaws; in this respect they resembled modern lampreys and hagfish, unloved parasitic forms that attach themselves to larger game fish such as lake trout. But the primitive fish probably weren't parasitic, since there was little for them to parasitize. Rather, they almost certainly swam mournfully along the bottom, sucking up debris and rotten organic matter into their jawless mouths. Rather an unprepossessing beginning for what was to become the crowning glory of evolution.

These early vertebrates quickly "discovered" the value of armor plating, probably to protect themselves from the marauding invertebrates. (In evolutionary terms, individuals having a greater tendency to develop bony armor had a selective advantage over those that did not, producing more successful offspring until eventually—read here: millions of years later—a large proportion of the population was protected in this way.) Armor made increased size possible, since the defended fish could now hold their own in undersea competition. At last, possessors of jaws found themselves at an advantage, and fish resembling our modern forms began to appear. Efficient, ferocious jaws may then have rendered the cumbersome external armor unnecessary or even a liability. In any event, among the resulting jawed fish was a peculiar side-group, unimportant in themselves, but destined to play a mighty role in the earth's future and our own.

Take a modern fish such as a trout and perch it on dry land: in the water it swam readily, largely by undulations of its body.

But on land these beautifully coordinated movements are useless and the poor animal flops helplessly about. In water, the fish "breathes" easily by taking water into its mouth, past its gills, and out again through slits on each side of the head. Oxygen from the water is transferred into the fish's blood, while carbon dioxide goes the other way. But in air this apparatus is useless and the stranded creature quickly suffocates, a fish out of water indeed. Unlike their aquatic relatives, land animals also generally have muscular limbs enabling them to move about, as well as blind-ended pouches (lungs) where oxygen/carbon dioxide exchange takes place. About 300 million years ago, a peculiar and seemingly unimportant side-group of fish (the "crossopterygians") differed from their cousins in having fins joined to their bodies with thick, muscular attachments that provided strength and flexibility of movement: just the prerequisites for walking on land. In addition, they had simple, balloonlike pouches enabling them to gulp air when this was necessary for them to survive. It must have tasted to them like a good breath of fresh water.

It is easy to think of these animals as bold pioneers, courageously setting forth on the world's greatest adventure: the conquest of the land. Actually, they were nothing of the sort. Living in freshwater ponds and swamps, many of the early crossopterygian fish must have died as the climate changed and their aquatic environment dried up. The survivors were those equipped with primitive lungs, enabling them to stay alive by gulping air during the parched, anguished times between cooling rains. (Similar forms persist today, in the lungfish of South America, Africa, and Australia, true fish that nonetheless regularly remain out of water for weeks at a time.) When times got hard those fish equipped with both lungs and fleshy fins—again, like our modern lungfish—could wriggle over the land in search of another pond. Most likely, they simply dragged their bellies through the mud, up and down drying river beds. If they had any thoughts at all, our fishy ancestors were doubtless yearning to recapture the only way of life they had known: a wet one. And natural selection was favoring those who were most adept at doing it. With a strongly conservative intent, one of life's most radical adventures had begun.

Since plants had preceded them onto land, the early terrestrial animals found a rich food source waiting for them, as well as a generally unexploited environment. It must have been a relative Eden, since most of the direct competition among animals was back in the ponds and swamps, and most of the thorns and poisons—subsequently evolved by plants as defense against animals—had not yet been called into existence. A group of animals thus luxuriated on the shore, as increasingly successful landed immigrants. However, they were not entirely adapted to life on land, since they still had to get their feet wet once again in order to reproduce: their soft, jellylike eggs, adapted to water rather than land, would otherwise dry up and perish in the hot sun. These were the amphibians, represented by the frogs, toads, and salamanders of today. It then remained for certain of these amphibians to break the chain binding them to the water by evolving a hard-shelled egg within which they could carry a protective watery environment far from pond, stream, or swamp. Enter, the reptiles. (In a similar way, the earliest multicellular animals had derived some independence from their environment by enclosing themselves and carrying around a substitute for sea water, something we now call blood.)

With their newfound independence, the early reptiles swarmed over the primitive landscape, evolving into a great diversity of forms, including the mighty dinosaurs.

All the while, early in this great adaptive radiation of reptiles, a small sidebranch developed once again, once again carrying with it a number of peculiar, distinguishing characteristics, traits that would have seemed innocuous enough at the time, but would ultimately be revealed as evolutionary blessings. For example, some of these reptiles had their elbows rotated backward and their knees forward, so that instead of suspending themselves between four pillars as present-day lizards do, their bodies were elevated directly above their arms and legs. Their teeth also became increasingly specialized, some for grinding (molars), some for stabbing (canines), and others for cutting (incisors). This dentist's delight reflected their capacity for a more wide-ranging, omnivorous diet. By contrast, the teeth of modern reptiles are monotonously similar to each other: a cautious glimpse into a crocodile's mouth should convince the skeptic.

The internal anatomy was further rearranged in this sidebranch, as it had been in the earlier transition from fish to amphibian, and again from amphibian to reptile. Reptilian scales gave way to insulating hair as these animals stumbled upon the trick of maintaining a constant internal body temperature in spite of the vagaries of the external environment. After all, air temperature changes much more than does water temperature. "Warm-bloodedness" therefore provided yet further independence from the outside, making these creatures more flexibly adapted to the land than their reptile ancestors had been. It also opened up regions of climatic extremes inaccessible to the dinosaurs and their relations. In addition, whereas the reptiles were limited to activity only at certain times of the day when the temperature was right, these new forms were free to design their own schedules.

These animals also made yet another major evolutionary discovery: developing young could be harbored within the mother and nourished by her blood supply. In this way, they were afforded greater protection and a constant temperature. And even after birth, nourishment was insured by feeding the young a fat- and protein-rich secretion produced by specialized sweat glands of the females: the mammary glands or breasts. Other important changes took place, such as building certain jaw bones in the middle ear, providing better hearing. These animals—the mammals—were initially rather small and unassuming, resembling modern shrews in appearance. They probably led fearful, inconspicuous lives in the shadow of the reptile giants, not unlike their vertebrate forebears in the primitive seas, perhaps occasionally snatching an unprotected egg. There was little opportunity for dramatic evolutionary expansion, since the dinosaurs and their reptile allies were already successful and flourishing, and there was little room left on the evolutionary stage while these monsters still held sway.

But then, about 100 million years ago and after an uncontested reign of more than 80 million years, the dinosaurs rapidly dwindled and disappeared. The causes of this massive die-off are unclear and still debated today. Perhaps those patiently gnawing mammals, with their superior reproductive methods and cunning ways, ate too many of the Goliath's eggs. More likely, the environment changed, leaving the dinosaurs too highly specialized to adapt. Maybe things became too dry, or too cold, and perhaps this was abetted by the impact of a large meteor, whose debris darkened the sky and cooled the ground for years on end.

In any event, the dinosaurs evidently lacked the flexibility necessary for long-term evolutionary survival. With their primary competitors eliminated, the field was thus opened for a riot of evolutionary experimentation among the early mammals. Flying, swimming, burrowing, grazing, browsing, drowsing, hopping, hanging, and predatory mammals filled the available niches; the surviving reptiles were restricted to relatively few representatives, concentrated in the tropics. Among the early mammals, one group took to the trees and retained many of the primitive features of their shrewlike precursors. These nondescript, apparently unspecialized little creatures deserve special watching, for among them—once again—are our ancestors.

Life in the trees exerted its own peculiar pressures. For one thing, it was necessary to move along branches rather than the flat ground. A grasping hand was more useful for this than a hoof or a claw, so these tree-dwelling forms—the primates—evolved long fingers and toes, with thumbs that could be opposed to them, providing a firm grip on cylindrical, three-dimensional objects. Thin, flat nails provided extra options without impeding the delicate work of the fingertips.

Further, most mammals had already become (and still are) "nose animals," guided about by a superbly developed sense of smell. But smells don't last long in the branches: treetops are windy places, with no continuous substrate on which a scent trail can be deposited or followed. So animals possessing a genetic tendency for a well-developed olfactory apparatus were simply wasting their time and energy, compared to others relying more on other senses, notably vision. The primates thus developed keen eyesight, particularly stereoscopic (three-dimensional) vision. With it, they could judge accurately the position of a branch or vine so as to grab it while leaping fifty feet or more above the forest floor. Stereoscopic vision, in turn, required that the eyes be rotated forward so they could focus independently on the same object, thus providing depth perception. This meant that the eyes had to lie in the same plane, like those of owls, rather than one on each side of the face, like dogs'. And this rotation required in turn that the doglike muzzle, still found in the terrestrial baboons for example, be greatly reduced in size, since it got in the way. Conveniently, our ancestors had pretty much given up being nose animals by this time anyway, so losing our muzzles was no great loss. Besides, our dextrous hands enabled us to explore the world manually, and to hold things up to be examined by our improving vision. The muzzle was thus expendable, along with the characteristic mammalian nose bristles. We became hand-eye animals.

Just when it looked like we would spend our future peacefully, if uneventfully, in the treetops, for some peculiar reason we decided to come down. This decision, in a sense, was no less momentous than the initial anonymous conquest of the land several hundred million years before. (It should be noted in passing that those expert tree climbers, the squirrels, descend a trunk head first, by adroitly rotating the bones of their wrists. Primates, by contrast, descend head up and rear end first. It is tempting to conclude that we have been backing into our future ever since.)

If our style was inauspicious, our motives—what biologists would call selective advantages—for resuming the terrestrial life were also unclear. Perhaps we had become too heavy to negotiate successfully the passage from tree to tree, since the weight-bearing capacity of branches decreases with distance from the trunk. We may have found ourselves forced to run from tree to tree along the ground. (But why, then, did we get so big in the first place?) Or perhaps our bold descent to the ground was but another would-be conservative venture that only appears daring in retrospect, like the first forays of the freshwater crossopterygan fish long before. Maybe the climate became increasingly dry, causing our forest home to thin out, gradually changing from lush jungle to sparse savannah as we find in much of east and south Africa today. Even now, the African savannah teems with wildlife, potential food for the omnivorous primate. Perhaps we were simply attracted to the better hunting down on the ground. On the other hand, and this notion is less pleasing to the human ego, perhaps we were forced down by other primates that were better adapted to an arboreal existence. Maybe some ancestor of the gorilla or chimpanzee, whose descendants today are losing out to the human presence in modern Africa, temporarily defeated us in the trees several million years ago, whereupon we moved reluctantly, in defeat and perhaps with no small resentment, to the flat, featureless, and seemingly marginal ground floor; refugees from the trees, pushed onto the plains.

Once established on the ground, we put our arboreal adaptations to good use. By flattening our feet and modifying our skeleton and leg musculature, we eventually stood straight. This upright stance may have caused certain large predatory animals to hesitate before attacking us. It may have given us a better view over the waving savannah grasses. But more important, it freed our arms and hands for other functions. Hands that had evolved for grasping branches could grasp them still, but to wield them as tools for digging up roots, or as weapons against would-be predators, prey, or each other.

The keen stereoscopic vision, so essential in the treetops, also permitted the precise hand-eye coordination that was needed for weapons to be used successfully. It is an important example of an early evolutionary teamwork, between the biology of what we *are* and the culture of what we *do*. When the team worked well together, the combination was unbeatable, as shown by the spectacular success of Homo sapiens in evolving and populating the planet.

Even with our biology and our newfound culture nicely complementing each other, we were still relative weaklings on the African savannah, still living precariously. So, those of us able to make efficient use of arms and hands were undoubtedly more successful in compensating for our physical weakness compared to the lion or the elephant. To a degree that may have never been true before in the history of life, we began to profit from what we *did* no less than from what we *were*. Digging, gathering and storing food, making rude shelters and clothing, killing other animals for food, as well as defending ourselves and our children from marauding beasts including other members of our own species: these things, as much as the structure of our lungs or our teeth, became a major arena for evolutionary success.

Not that our anatomy didn't continue to evolve. Bipedalism—walking upright on the hind legs—may have been, appropriately enough, a crucial first "step" toward becoming human, since it helped free our hands for tool use, tool making, and communication, rather than simple locomotion as in the other mammals. We quickly exploited an ability to adjust and improvise, traits that served a weak-bodied, large-brained animal so well in its new savannah home. Something crucial had begun: although biological evolution continued (just as it continues today), for the first time in the history of life, a new evolutionary factor had become significant, a factor whose importance was to challenge and eventually to overtake our biology. We started to become cultural animals.

The stage of human evolution described here is represented by a series of fossils known as Australopithecines, "southern apes," from south and east Africa. These early members of the human family were represented by *Australopithecus afarensis*, the earliest of which lived about four million years ago. They walked upright, were about four feet tall, and had brains that were about one pint in volume; approximately the size of modern chimpanzees'. In another two million years, there were at least three species, two heavy boned, slow moving, strictly vegetarian types (*Australopithecus robustus* and *A. boisei*), both of which apparently went extinct without leaving significant descendants, and a more slightly built, omnivorous form to which all of us trace our ancestry. Although we have no way of telling what these man-apes looked like (whether they were covered with hair like modern apes or virtually naked like us, for example) certain characteristics can be determined from their skeletons.

They stood somewhat less than five feet tall when adult, weighed less than a hundred pounds, and walked fully erect on their hind limbs. They had receding chins and foreheads, prominent eyebrow ridges, and teeth that look strikingly human. They also made simple chiseled tools and had brains that were nearly twice as large as those of the most primitive Australopithecines, roughly halfway between chimpanzees' and modern *Homo sapiens'*. They are the first representatives of the genus *Homo*, and have been given the species name *habilis*; *Homo habilis* in English would be something like "handyman."

Many anthropologists now agree that the early Australopithecines were meat eaters—if not exclusively so, at least for a large portion of their diet. And *Homo habilis* was literally "handy" as well, using his hands to wield tools and weapons. Skulls of medium and large animals, apparently killed by hard objects, have been unearthed from caves where *Homo habilis* lived and ate. The tree-dwelling crossopterygian fish had come once again to the land, and once again, they were making good.

We are justly proud of our enormous brain. Throughout the Australopithecine stage, however, that brain, although large by most animal standards, was still nothing to get excited about: no larger than a modern gorilla's. But beginning with those early, savannah dwelling, hunter/gatherer/scavengers, our brain enlarged at a fantastic pace, attaining human proportions (roughly twice the volume of a gorilla's) in a mere two million years. Although this may seem like a terribly long time, in evolutionary terms it is astonishingly short. For the brain to double in size within that timespan bespeaks the action of enormously powerful selective forces. The African man-apes were clearly living in an environment that placed a special premium on brainpower: contrary to what is commonly thought, our use of tools and weapons, and indeed, our remarkable development of nonbiological (cultural) extensions of our bodies, was not the *result* of a large brain, but rather, its *cause*. The gorilla-brained Australopithecines found themselves fortuitously preadapted to primitive tool use by virtue of their arboreal past. The cleverest of them were most successful in using tools and weapons to obtain food, to triumph over their enemies, and to support their families. Such individuals left more offspring. After all, a naked ape on the African savannah had to have something going for it in order to be successful; given our relatively weak backs, evolution strongly favored those of us with strong minds.

But the big-brained animal we call ourselves did not evolve solely in response to tool use. Other pressures were operating, all pushing in the same direction. Communication between individuals was of great advantage, whether to coordinate a hunt, plan a berry-picking, nut-gathering, or root-digging expedition, arrange for unified group defense, describe the locations of prey, enemies, water holes, or prospective shelter, consider alternative courses of action, or instruct offspring in the increasingly complex skills necessary for successful living. With the development of language, enormous horizons opened: we could discuss options, employ abstractions, evaluate the past, admire the present, plan the future. Individuals possessing these capacities enjoyed tremendous advantages, and so the evolution of the human brain received another forward thrust.

In addition to tool use, it therefore seems likely that natural selection smiled upon the early ability to interact effectively with others. This received additional impetus *from* brainpower, and in turn, provided additional pressure *for* the evolution of yet brainier specimens, especially those who could "handily" employ other individuals as "tools" for their own success.

Tool use, communication, and brain evolution thus interacted to produce a positive feedback ("vicious circle") loop. The ability to utilize tools and to coordinate our activities exerted initial selective pressures for increased brain capacity. This in turn made increased tool use and interpersonal collaboration possible, while the availability of increasingly sophisticated devices and opportunities further enhanced the desirability of a large brain. When *Homo habilis* evolved into *Homo erectus* ("Peking man"), about 1.5 million years ago, this process was well under way. By 350,000 years before Christ, *H. erectus* was using fire and ocher pigments, and by 60,000 years ago, *Homo sapiens* was on stage. Anthropologists are divided over whether the Neanderthal people were among our direct ancestors, or simply distant cousins. But it is generally acknowledged that *Homo sapiens neanderthalensis* were human beings, just a different subspecies or race from the modern form, *Homo sapiens sapiens.* The Neanderthalers buried their dead with flowers and almost certainly engaged in primitive religious rites. Archaeologists have found the skulls of great cave bears, enshrined by the Neanderthalers on rudimentary altars; something rather interesting must have been going on within the Neanderthalers' skulls as well. Cro-Magnon people were making beautiful paintings on cave walls, as well as carved figurines, 25,000 years ago, by which time, biologically, we were much the same as we are now.

Biological evolution is the most fundamental force for change acting on human beings; it is also the slowest, the most resistant, and perhaps as a result, the one most likely to be ignored. It is our species-wide purloined letter, responsible for those aspects of our bodies and our behavior that are so widespread and familiar that hardly anyone notices.

The long climb out of the organic soup had ended; or rather, our biology had produced what we are today, what we see when we look in the mirror. But to some extent it has been out of the soup and into the fire. Twenty-five thousand years ago, our metaphoric tortoise had pretty much reached where it is right now; human biology had arrived. But the human experience, by contrast, was just beginning. The last few steps (especially since we descended from the trees) may have been somewhat faster than they were before, but biological evolution has never been known for its speed . . . at least, not compared with its cultural cousin. Imagine a tortoise running (or rather, plodding) a marathon: virtually that entire route, from the origin of life to Cro-Magnon times—the long, winding journey through the jawless fish to the amphibians, the reptiles, the early mammals, and so forth—would be equivalent to all but the last foot or so. The last little bit of the tortoise's journey, from the Cro-Magnon days to the 1980s, would all be compressed into just a tiny fraction of the trip. But during that brief interval, while the tortoise takes yet another laborious step, an awful lot has been happening.

CHAPTER THREE

ANATOMY OF THE HARE

God took man as a creature of indeterminate nature, and, assigning him a place in the middle of the world, addressed him thus: "Neither a fixed body nor a form that is peculiar to thyself have we given thee, Adam; to the end that according to thy longing and according to thy judgment thou mayest have and possess what abode, what form, and what functions thou shalt desire. The nature of all other things is limited and constrained within the bounds of laws prescribed by us. Thou, constrained by no limits . . . shalt ordain for thyself the limits of thy nature . . . As the maker and molder of thyself in whatever shape thou shalt prefer, thou shalt have the power to degenerate into lower forms of life, which are brutish. Thou shalt also have the power, out of thy soul and judgment, to be reborn into the higher forms, which are divine.

—COUNT GIOVANNI PICO DELLA MIRANDOLA,
Italian Renaissance humanist, in
De Dignitate Hominis
(On the Dignity of Man)

Maker and molder of ourselves; constrained by no limits; choosing by our own judgment what form and function shall be ours: in short, God—according to Pico—bequeathed us cultural evolution. According to evolutionary biologists, the process worked somewhat differently: the oversized human brain, combined with hands capable of dextrous manipulation, helped us achieve ways of living and doing that went far beyond our bodies, and which very quickly overwhelmed the pace of biological change. Either way you look at it, early in human history we came under the influence of cultural evolution as well as biological evolution. And whereas sometimes the two operate harmoniously, sometimes they do not.

To some people, the word "culture" summons images of symphonies and painting, poetry and the Kennedy Center, opera and the weightier aspects of educational television. For our purposes, however, culture is much more general, widespread, and essential. Whereas human beings could survive without chamber music, they could not survive without culture; they could not survive without those various extensions of their bodies that they universally create, and without organizing themselves and orchestrating their lives in meaningful, symbolic, and profoundly extrabiologic ways. The culture of a particular human group might be defined in general terms as the sum total of all ways of living practiced by that group. The British anthropologist Edward Burnett Tylor first suggested this more inclusive concept in his book *Primitive Culture*, published in 1871. Tylor defined culture as "that complex whole which includes knowledge, belief, art, morals, law, custom and any other capabilities and habits acquired by man as a member of society."

Some major cultural inventions of the human species include tools and weapons, fire, agriculture, animal domestication, cities, smelting of iron, copper, and other metals, the wheel, gunpowder, the compass, the steam engine, the internal combustion engine, the airplane, penicillin, computers, and atomic energy, not to mention peanut butter, Hula Hoops, and underarm deodorants. Other major cultural developments, less likely to leave physical artifacts, but no less important to the human story, include language, writing, religion, political systems, laws, patterns of economic exchange, and science. Of course, many more factors could be added; to make the list complete, nearly every complex detail of human behavior would have to be included.

Cultures have progressed in different ways among different groups. Although certain phenomena may be universal—such as language, tools, and the recognition of some patterns of kinship relations among individuals—others are limited to specific local populations. The wheel, for example, was essentially unknown to New World inhabitants before the arrival of Europeans (although, interestingly, wheeled children's toys have been found among Aztec artifacts), while the practice of mate swapping, long characteristic of Eskimos, only recently appeared in certain segments of modern American culture, and less than ten years later, it seems to have disappeared again.

REGARDLESS of its details, however, all human culture shares something essential, a crucial characteristic that distinguishes its evolution from the slow, plodding biological evolution that preceded it: cultural evolution is independent of changes in genetic makeup. Whereas every individual is the *product* of biological evolution, he or she cannot evolve biologically. On the other hand, each of us can experience a variety of cultures during a single lifetime, and not only by visiting them. An aborigine brought from central Australia to western Europe can leap hundreds if not thousands of generations of cultural evolution within a few years. But the pace of cultural evolution is such that even the sedentary human being, just by living a few decades, is guaranteed to see a remarkable array of cultural practices come and go. New customs and inventions can spread from people to people, like lightning. Just like conductors of electricity, in fact, people can transmit currents of culture, as quickly as they are generated. But in another parallel to conductors of electricity, the people themselves will not really be changed, any more than a copper wire "evolves" during normal use.

This disparity between people, biologically, and people, culturally, occurs because culture is transmitted by a new mechanism, in which individuals may directly acquire characteristics that were initially obtained by another. It is a new process, quite divorced from normal biological events. With elaborate and sophisticated new technology, we can sometimes introduce the genes of one living thing into another; left to our own devices, this does not happen. Once formed as fertilized eggs, we are impenetrable to each other's genes. But once formed as human beings, we are readily penetrated by each other's ideas and cultures.

Whereas biological evolution is now acknowledged to proceed largely by natural selection (Darwinism), cultural evolution progresses by Lamarckism, "the inheritance of acquired characteristics." Herein lies the crucial difference between the tortoise and the hare. First proposed by Jean Baptiste Pierre Antoine de Monet, Chevalier de Lamarck, notably in his *Philosophie Zoologique* (1809), Lamarckism has since been generally discredited as a mechanism for biological evolution. But when it comes to culture, Lamarck was on the mark. His theory was based in large part on the effects of use and disuse: when a structure is used, it often grows, when ignored it atrophies. If these changes are then transmitted to one's offspring, a mechanism for long-term change is available, at least in theory.

Lamarckism suggests, for example, that the long necks of giraffes evolved because ancestral generations of giraffes kept stretching their necks, trying to reach food high in the African treetops. Similarly, a weight lifter could insure large biceps for his offspring simply by developing his own muscles. Unfortunately for Lamarck's reputation among biologists, the living world just doesn't work this way; there appears to be an almost complete separation between the somatic tissues of a living creature (its body) and its genes. Information flows from the DNA to the proteins, from genes to bodies, but not vice versa. Thus, biologists have cut the tails off many generations of adult mice, only to find each new generation is no more tailless than the ones before. And after thousands of years of circumcision, Jewish boys are still born with foreskins. Biological evolution depends on changes in *genes*, and genes are not changed by experiences acquired during the lifetime of the individual.

But culture is different. When medieval Japan was introduced to Western ideas and technology in the latter decades of the nineteenth century, it was able to absorb much of that culture with incredible speed. Enough to defeat Russia in the Russo-Japanese War and immediately establish itself as a major world power in the twentieth century. Rather than waiting for mutation and recombination to blunder onto the appropriate characteristics (even with the impossible assumption that such traits could be produced by biological evolution) and then waiting for their advantage to be recognized by natural selection, the Japanese people could immediately swallow whatever aspects of Western culture they considered desirable—just as the Russian people accepted the precepts of Marx and Lenin only a few decades later, and transformed their culture with equal speed. All this in less than one generation.

Similarly, when Alexander Graham Bell invented the telephone, it rapidly spread throughout the population because of its great usefulness. Imagine a biological evolutionary invention of equal value: relying upon genetic mechanisms of formation and transmission, it would spread much more slowly. Even the extraordinarily rapid evolution of human brain size, for example, proceeded at less than a snail's pace compared with the evolution of the telephone. If using a telephone depended on the existence of telephone genes, then Bell's children might possess that capacity (assuming it was a dominant trait), and perhaps 150 direct descendants would currently be using telephones. Instead, the technique was transferred by cultural evolution, and several billion people are telephone users today. Moreover, the external shape and internal design of the telephone has also been radically modified. During this time, essentially no biological evolution has occurred.

In addition, biological evolutionary change depends as we have seen on the gradual accumulation of many small steps. This is because a gross rearrangement of an organism's genetic system might well kill it, or at least cause so much disruption that it would be at a strong selective disadvantage. And the larger this disruption, the greater the impact. Biological evolution is thus forced to proceed by small steps because each step is essentially random. Natural selection sifts and chooses from the array of variations available to it, so that given enough time, order appears out of potential chaos.

The circumstances surrounding the emergence of cultural innovations are actually less well known than their biological counterparts. Like biological innovations, cultural innovations sometimes seem to appear randomly; it is almost impossible to predict the debut of human discoveries, innovations, or changes in customs. And who can describe the events that precede and give rise to an original idea? Like biological mutation, certain environments seem to stimulate more cultural innovation than others. The Renaissance was a "mutagenic" cultural environment, for example, whereas by contrast, the thousand years or so preceding it were not.

Whatever the causes of cultural innovations, they can spread not only independent of our genes, but in response to conscious decisions. When Europeans first tasted Oriental and Caribbean spices, for example, they were enchanted and wanted more. New trade routes were opened, new cities and a new merchant class became powerful, and human society—as well as human cooking technique—was changed in just a few decades. The Old World, similarly, incorporated tomatoes, potatoes, and corn from the New World, just as the New World quickly incorporated gunpowder and horses from the Old. Cultural diffusion can even outrun the actual migration of people themselves: a tiny band of Spanish adventurers conquered the much more numerous (and in many ways, more sophisticated) South American civilization largely because horses were unknown to the Incas and the Aztecs. A few hundred years later, the white conquerors of North America encountered

Plains Indians who were expert horsemen, but who had never met a European.

The spread of Valley Girl talk, skateboards, and Hamburger Helper may appear to be almost random and undirected, but in fact, there is often method to such apparent madness; new cultural patterns can be incorporated by the direct design and intent of human beings ("Mr. Watson, come here, I need you!") rather than under the control of a blind, opportunistic natural process. And whereas some such innovations can be trivial, others can entail major reorganizations of the entire social, technological, and personal fabric, for better or worse. As Leibnitz put it, *nature non facit saltus* (nature does not make leaps). But culture does. Sometimes it tries to look before it leaps, and this is another difference between cultural and biological evolution, since the latter is "blindly opportunistic," as the great geneticist Theodosius Dobzhansky pointed out. At other times, cultural evolution seems no more preplanned than its biological counterpart. Unlike nature, *cultur facit saltus.*

Human beings have developed Hinduism, the Renaissance, Christianity, the Industrial and Scientific Revolutions, Victorianism, New Deals, Five-Year Plans, Democracy, Communism, nuclear weapons, TV dinners, X-rated movies, video games, and birthday parties. Cultures may add, delete, or exchange major (or minor) components, transforming themselves within a fraction of a generation, as occurred in turn-of-the-century Japan and indeed, most cultures in contact with modern Western technology. Nearly everyone in the world has heard a transistor radio, although most do not know what a transistor is, and Burmese farmers often listen to rock and roll while plowing their fields by buffalo. In Ecuador, bananas are piled by hand into dugout canoes, as they probably have been for thousands of years, and then off they go, powered by outboard motors. Sometimes, as in these cases, one culture borrows traits from another. Often, the exchange is mutual, as with the fur parkas, snowshoes, and igloo design adopted by Americans and Europeans from the Eskimos, and the rifles and snowmobiles adopted, in turn, by modern Eskimos. In other cases, aspects of one culture are forcibly inserted into another, as with missionary Christianity or Islam.

The nature of cultures can be debated endlessly; they can be considered to be large, complex, and interdependent systems, like organisms, or the product of numerous discrete units analogous to genes in biological systems. Whereas cultures may undergo gradual transition, especially as a result of new environmental challenges (such as the Ice Age, the spread of industrialization, or resource shortages), they are also capable of gross mutation. The frequency of such convulsive change has been increasing in recent years, reaching a torrid pace during this century. And it shows every sign of gathering even more speed in the future.

Having distinguished biological from cultural evolution, we might further subdivide cultural evolution into two major components: social evolution and technological evolution. Social evolution includes forms of law and government, economics, the basic structuring of nation, family, work, and environment, music, art, literature, religion. Social evolution occurs in a time frame of decades, more often hundreds or even thousands of years. Although very fast and flexible compared with biological evolution, social evolution is nonetheless very slow compared with the other great pillar of cultural evolution, technological change. Of the two, technological evolution is by far the most rapid, bringing forth innovation at a rate that would be staggering for societal change, and unimaginable for biological change.

Thus, societies evolve with amazing speed by standards of natural selection, but slowly by standards of technology. Compare life in the United States today with 150 years ago, for example. The technology is dramatically different, including electrical appliances, automobiles, medicine, and nuclear weapons, to mention just a few. The societies, by contrast, are also different, but not wildly so: there may be fewer biparental households, but there are still households; the work week may be shorter, but there is still a work week; the government may be larger, but it is still a recognizable government, indeed, basically the same government—even the same Constitution—as existed in Andrew Jackson's time. By contrast to the revolutionary changes in technology and the moderate changes in society, the American of the Mexican-American War would be biologically indistinguishable from the American of today.

Technology proceeds by discovering and implementing methods by which human beings can act on the inanimate universe—on metal and stone, plastics and electronics, rocket ships and microwave ovens—or on the living, but nonhuman world: the planting and breeding of crops, the maintenance or exploitation of other animals, the care of our own innards. Technological evolution follows scientific discovery, and the basic insights of mathematics, physics, chemistry, and biology. It uncovers the laws of nature, and seeks to manipulate them; but, unlike social evolution, it does not write these laws.

Technology, for better or worse, has been progressing. It has allowed us to manipulate more of the world, and with less effort, than ever before. In its own way, biological evolution also involves progress of a sort, notably greater independence from one's environment (note the "progression" from naked genes, to bodies, to semiterrestrial amphibians, to warm-blooded mammals, etc.). Biological evolution also involves the accumulation of increasingly precise adaptations, such that highly evolved living things tend to "fit" their environment rather closely. However, there is very little evidence for progress in social evolution. Democracy, as practiced by the Greeks, has not been noticeably improved upon 3000 years later. Experiments with utopian societies come and go, with almost depressing regularity, and dictatorship—often with religious underpinnings as in the Ayatollah's Iran, or based upon secular theologies such as in the Soviet Union, are at least as old as the Age of the Pharaohs. Unlike technology or biology, societies do not so much evolve as revolve.

Are modern philosophers and theologians any farther advanced than Aristotle, Plato, Gautama Buddha, or Christ? Are today's Pulitzer or Nobel Prize winners better than Homer or Chaucer? Slavery is legally abolished worldwide, but forms of serfdom still exist and apartheid isn't very far behind (or ahead). Royalty and other forms of totalitarianism come and go, as does democracy. Religion oscillates from fundamentalism and absolutism to liberalism and "secular humanism" under various guises. Nontechnological, social culture may be characterized by circularity, or perhaps perturbations from a baseline, with repetitive reversions to that baseline, followed by deviations in the opposite direction; *Plus ça change, plus c'est la même chose* (the more things change, the more they stay the same). Perhaps the changes that characterize social evolution are due less to progress as such, than to the restless intelligence, interpersonal and intersocietal competition, and random mood swings writ large, which combine to make nontechnical social traits unstable or at least with a stability that seldom exceeds several hundred years. Thousand-Year reichs, of any sort, are rare indeed—and perhaps blessedly so.

It seems most likely that social evolution is characteristically unstable and unprogressive in large part because it involves the relationship of human beings to other human beings. The laws of nature can be revealed, but never repealed; the laws of human society, by contrast, are ours to write, erase, and rewrite. We can violate the latter, never the former. And we are much more likely to "get somewhere" with technological evolution since it is nearly always faster than social evolution.

We can better understand the differential pace of biological, societal, and technological evolution by considering those factors that *impede* change. Technological change is limited mainly by ideas, and also, by physical opportunity: access to the necessary materials, capital, support systems, etc. Societal change, by contrast, is limited by the presence of human beings themselves, which act like a buffer in a chemical solution. Thus, we may innovate, copy, and modify styles of clothing, for example, but human beings—by their very nature, it seems—insist on having clothing of some sort. Biological change, finally, is the most sluggish of all, since here the ballast is provided not by ideas, or even by human values, attitudes, and preferences, but rather by the human gene pool itself, which is the least modifiable of all human traits.

Much of the discordance in human existence may be due to the conflict between technological evolution and social evolution, both of these processes interacting within the cultural sphere; this particular conflict increasingly occupies sociologists, anthropologists, and specialists in the social management of technology. It well deserves such attention. For our purposes, however, it shall be useful to combine all fundamentally nonbiological factors (including both social and technological evolution) under a single rubric, and to designate them as "cultural evolution" as opposed to "biological evolution." In the concord and discord between these two, we shall search for the roots of ourselves, that which makes us human, happy, magnificent, and miserable.

These alternative levels of human evolution are similar to the anatomical distinction between our higher, cerebral functioning and lower, more primitive brain systems. Dr. Paul D. MacLean, former head of the Laboratory of Brain Evolution and Behavior of the National Institute of Mental Health, has emphasized a three-tiered conception of the human brain:

> Man finds himself in the predicament that Nature has endowed him essentially with three brains which, despite great differences in structure, must function together and communicate with one another. The oldest of these brains is basically reptilian. The second had been inherited from the lower mammals, and the third is a late mammalian development, which . . . has made man peculiarly man. Speaking allegorically of these three brains within a brain, we might imagine that when the psychiatrist bids the patient to lie on the couch, he is asking him to stretch out alongside a horse and a crocodile.

The distinction between cultural and biological evolution is not quite the same as that between the cerebrum on the one hand, and the paleomammalian and reptilian remnants on the other. Rather, it is between the entire human brain, in all its biological parts, and the most conspicuous *product* of that brain: culture.

Even this distinction is somewhat arbitrary, since as we have already seen, human biological and cultural evolution have been intimately connected, mutually supportive, and in a sense, inextricable. We are cultural animals. Culture is as "natural" to Homo sapiens as hooves to a horse or scales to a crocodile. A human being without culture would be as bizarre as a naked peacock, or a porcupine without quills.

In fact, culture is not even unique to our species. Chickadees in Britain, for example, quickly learned that they could sip the cream from unhomogenized milk by removing the foil caps; a tradition developed, with birds following milkmen and descending on the bottles before the homeowners did. Rats avoid poisoned bait, and red squirrels learn how to open acorns, once again without biological evolution taking place. Habitual methods of food preparation have been innovated by a single adult female Japanese macaque monkey, from whom they spread in a truly Lamarckian fashion to other troop members. She developed the now-famous tradition of dipping sweet potatoes in the ocean, to salt them; another time, the same individual discovered that she could separate wheat from sand by dropping handfuls into a stream, whereupon the wheat would float and could easily be scooped up. Other Japanese macaques picked up the behavior, without their genes changing in the slightest.

This renowned spread of new cultural traits among Japanese monkeys has even given rise to a new, culturally elaborated myth, the so-called "hundredth monkey phenomenon." According to this tale, once the hundredth monkey does a particular thing, the trait in question spreads rapidly throughout the population; this story is often told in the hope of encouraging the average citizen to become politically active, to "become the hundredth monkey." Although the hundredth monkey story is merely a well-intentioned fabrication, it is useful for our purposes, exemplifying another level of cultural evolution: the cultural evolution of an account of cultural evolution. (Mathematically inclined readers might call it "cultural evolution squared.")

When human beings invented tools and conceptual language, discovered the uses of fire and of domesticated animals, these were gradual—even slow and tedious—processes by modern standards. And in a sense, they were still at least minimally concordant with our biological evolution. But as the process has continued and the pace of cultural evolution has increased exponentially, the connectedness with biological evolution has grown more and more tenuous. A human being can walk to a bicycle, then pedal to an automobile, then get on a jet plane, or even a rocket ship without at any one point definitively severing his or her connectedness from the shambling bear or the sprinting cheetah. But the transition is not really a seamless web. When the quantitative difference becomes great enough, a qualitative difference has effectively been reached, even though there is no clear point of transition: a cheetah could fly as a passenger to the moon, and an astronaut stalk gazelles on the African savannah, but each would nonetheless be out of place. Yet, we still get occasional, nearly pathetic reminders of our animalness. There is something almost comic, for instance, about high-tech spacesuits designed with little tubes permitting primitive, smelly organic fluids to pass in and out, and even about the suits themselves, with their ungainly forked-stick shape, necessary to accommodate the two-legged space voyager who is still the same bipedal primate who once-upon-a-time walked about on the African savannah.

⁂

CULTURE and biology need not always be opposed. Indeed, it may be that most cultural practices are either neutral with regard to biological significance, or even strongly adaptive, such as the tenacious cultural defense of some form of marital bond, or indeed, most things that human beings do in their daily lives. This is not surprising, since cultures themselves succeed each other just like populations or species. As we have seen, a process analogous to natural selection determines their spread, although the mechanism is of course quite different. Natural selection does not involve any conscious choice on the part of those genes and gene combinations that experience maximum reproductive success. By contrast, culture often proceeds by the intentional selection of specific practices from among a large array of those available. In this sense, then, it is "teleological" or goal directed in a way that biological evolution never is. For example, Marxist ideologists have been able to design and carry out particular cultural practices in many nations within the past fifty years, just as capitalist ideologists have done for several hundred years before that. Although the results may not always have been exactly as intended, and much of human cultural evolution (like much of biological evolution) may be essentially random, the impact of conscious, end-directed choice in human cultural affairs certainly cannot be denied.

It may be argued that most human beings throughout most of human history have never had any real options for choice of cultural practices. Thus, the culture in which one is raised has a distinct and often pervasive influence, predisposing one's future decisions, and typically ruling others out altogether. Even if we were entirely free to decide upon our own cultural path, independent of our childhood experiences, the outlook of most people has always been narrowly restricted by the cultural models available to them. We simply do not have much chance to achieve drastic societal or cultural change. (Although, of course, we have even less chance to achieve biological change, radical or not.) Just as biological evolution must await the production of genetic diversity by mutation and sexual recombination, cultural evolution requires both exposure to new cultural practices, either through invention or observation, and the ability to adopt them.

For thousands of years, even while it was highly changeable

compared to our biology, human culture nonetheless changed slowly by modern standards. But then things began speeding up. The scientific revolution in particular began feeding upon itself, producing innovation at a greater and greater pace, while improved communication and transportation techniques enabled rapid dissemination of cultural change. Like when an asexual species suddenly becomes sexual, whole new avenues of (cultural) evolution were opened up and we have been rushing along them ever since, like a hare outrunning a tortoise.

Genetic combinations favored by natural selection are almost always adaptive; that is, they represent characteristics that tend to help their possessors do a better job of living and reproducing. If they didn't, they wouldn't be selected for. (There are some exceptions, as when a disadvantageous characteristic is naturally selected because it is somehow linked to another, advantageous one, whose benefit outweighs the other's detriment. But such events appear to be rare.) By contrast, success in cultural evolution is only indirectly determined by adaptive advantage. Thus, human culture may sometimes spread by physical superiority, as with Genghis Khan overrunning Asia or the white man overpowering the red in both North and South America. (The "physical superiority" here is of cultures, not necessarily of individuals; Genghis Khan's success was in fact due largely to the invention of stirrups, and the white man's, to horses and gunpowder.) But as mentioned earlier, success in physical combat is only one small aspect of total fitness, and conquering armies have often found themselves adaptively inferior to local populations in terms of their ability to deal with local conditions. The conquerors may thus absorb much of the "defeated" culture, coming to resemble it more than their own. In such cases, the losers may indeed have been beaten as individuals, but some aspects of their culture can still be victorious.

Biological hybridization is the union of two individuals from substantially different gene pools, usually members of different species. Sometimes when the hybridizers are similar but not too similar, the offspring show "hybrid vigor," and are hardier than either parent. When hybrids are made between different species, however, the results are usually adaptively inferior, since the combinations occur essentially at random, and each genetic system is adapted to itself, not to others. For example, given two well-constructed automobiles of different makes, it is unlikely that an unthinking exchange of parts between them would produce a better product. But if that exchange is performed by a competent mechanic, who consciously selects the best spark plugs while perhaps leaving the same carburetor, a real hot rod might result. Similarly with an exchange between cultures. Instead of the unplanned mixing characteristic of genetic hybridization, cultural hybridization often results in selection of the most appropriate combinations from both systems, leading to a more viable product than either was separately.

A successful culture may be superior in only a limited number of ways, as for example the recent successes of Western technological culture. The last 200 years have seen a drastic worldwide decline in the diversity of local cultures as "Westernization" has circled the globe. It is unlikely that this indicates true adaptive superiority. More likely, it reflects the pervasive influence of our superficially impressive and (perhaps superficially) effective science and technology. Only time will tell whether this system is adaptive over the long haul.

Culture may also spread by appeal to higher mental faculties, especially in combination with ascendant technology.* The rapid dissemination of Islam was facilitated by the former, and of missionary Christianity, the latter. Once again, the extent to which cultural practices are adaptive in the evolutionary sense can only be told by some historians of the distant future. Intellectual or

*In some cases, superior technology as such was not the major factor leading to the success of one culture over another. For example, missionary Christianity owed much of its successful spread to the fact that Caucasians introduced European diseases such as measles and smallpox, to which native populations lacked immunity.

ideological appeal has never been a direct factor in biological evolution, and such appeal may have no bearing whatever upon true adaptive value. On the other hand, an argument can be made that at least to a limited extent, cultural values are given emotional salience in proportion as they meet at least some of our biological needs. Success in evolutionary terms must be measured by continued existence over time; the absence of future humans to pass judgment upon a culture—or the entire species—will be ample judgment in itself.

One of the pitfalls of biological evolutionary change is that its reliance upon immediate adaptive advantage necessitates a rather shortsighted mechanism. Thus, short-term success may be achieved at the cost of long-term inflexibility and ultimate extinction. For example, an environment that is homogeneously light colored—such as occurs in White Sands National Monument, in New Mexico—selects for light-colored rodents, which are camouflaged from their predators. But when and if the background environment becomes darker, the light-colored mice—successful in the earlier environment—will quickly be selected against. Human cultures may also be susceptible to this capacity for overspecialization, finding themselves unable to adjust to changing conditions. They may achieve enduring success by anticipating, modifying, and compensating for the contingencies of the future. Or they may reveal that our unique teleological capacities are to us like the large canines were to the saber-tooth: useful for a time, but no ultimate security and maybe even a liability.

Unlike modern technology, some cultural practices have apparently existed in one form or another for thousands of years. They have demonstrated their adaptive utility and would presumably merit the imprimatur of biological evolution. For example, the Mosaic prohibition against eating pork may well have been prompted by recognition of the dangers of trichinosis. Whatever its theological underpinnings, it is thus good biology. Ironically, however, such prohibitions have in recent years been rendered largely unnecessary by further cultural practices such as hygienic hog-raising and careful cooking. Similarly, the east Asian and African practice of agricultural fertilization with human feces is poor biology because of the role of fecal matter in disease transmission, but good ecology in regions where soil is poor, scarce, or overused due to excessive human population.

In the absence of agricultural and culinary technology, and given sufficient time, one might expect the prohibition against pig-eating to become genetically incorporated within the human species, since those who refrained would have a lesser tendency of contracting disease and hence, a selective advantage over those that did not. Eventually—perhaps after 10,000 years or so—selection should therefore favor a genetic tendency to avoid pork, rendering cultural prohibitions unnecessary. The use of night soil fertilizers would ultimately depend on the balance struck in each local population between the abundance of disease organisms and the ecological demands on arable land. In both these cases, however, cultural evolution has overwhelmed whatever potential biology may have possessed.

Take another example: drinking milk. For the great majority of the world's human beings, it would seem as bizarre for an adult to drink the milk of a cow as it seems to us when the Masai drink the blood of their livestock. Most adult human beings lack the enzyme lactase, needed to digest the milk sugar, lactose. In fact, there is some correlation between the genetic ability to manufacture lactase and the complex cultural practices of dairying. Does this mean that there is a gene "for" dairying? Almost certainly not. But it does suggest that cultural practices—even complex and highly ritualistic or technologic practices—may be ultimately connected to our biology, often in unexpected ways.

It is extremely difficult to identify the degree to which human behaviors are genetically influenced, and to some extent, it is foolish to try. If a person stands five feet tall, it is meaningless to suggest that three feet and four inches of her stature is due to her genetic makeup, with the remaining one foot and eight inches due to her environment (nutrition, health, etc.). Rather, every inch of her height results from the interaction of her genes and her experiences. Certain human behaviors are nonetheless more rigidly constrained by genotype, while others are more flexible. Behavior that appears almost universal to all cultures, that has not varied during recorded history, and that is clearly of biological advantage, would seem likely to have genetic underpinnings. It is among such behavior that possible conflict between our biologically given inclinations and our culturally generated realities may be found.

The biological evolution of Homo sapiens has not stopped, even today. Nonetheless, it was virtually complete—that is, the human being we know today had already arrived—when human culture was really just beginning. Although the interval between the domestication of fire, for example, and the discovery of agriculture (perhaps 375,000 years) seems like an eternity to modern people, in fact it was a very brief time as biological evolution reckons these things. And whereas the interval between Gutenberg's invention of the printing press and Marconi's invention of the radio (about 500 years) is almost literally insignificant in biological time, it is immense in cultural time. Choose any 500 years of human history prior to about 1000 B.C.: although the cultural changes occurring during those years are bound to have been immense compared to the biological changes during the same time, such a timespan would be uneventful compared to the 500 years that our species has just witnessed.

The reason is the extraordinary pace of cultural evolution, freed from its biologic bonds. There is a general pattern for the rate of cultural change, differing in detail but generally accurate for most systems: cultural change has become approximately exponential. That is, the rate of change has itself been increasing, like a falling object accelerating by gravity, or—more to the point—like a rocket taking off. The general shape of human cultural evolution is therefore like this, ever-accelerating but becoming steeper as we approach modern times:

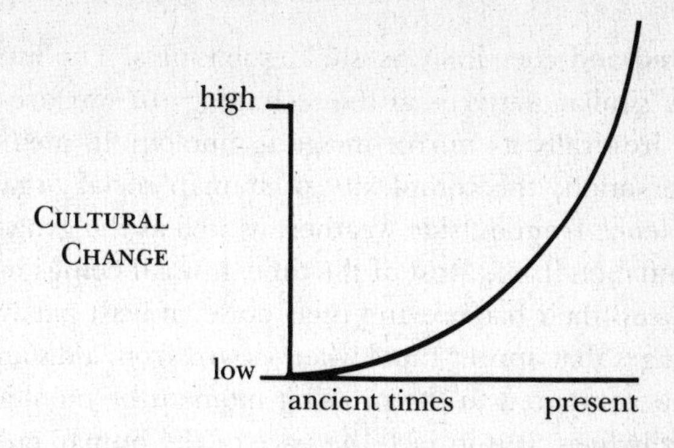

As just one example, let us consider power, defined by physicists as the ability to do work. For hundreds of thousands, or even millions of years, human power was limited by our own muscles. With the domestication of animals, perhaps 6000 years ago, we acquired ox-, camel-, and horsepower, as well as the basic simple machines such as the lever, inclined plane, wedge, wheel and axle, and screw. Then, we began harnessing the wind and the water, via sailboats, waterwheels, and windmills. Chemical power was unlocked after that, through the discovery of gunpowder. Electricity and the steam engine came on the scene only last century, followed shortly by the internal combustion engine and, just a few decades ago, the splitting of the atom with its release of colossal, unimaginable energy. More and more energy has been harnessed, in time intervals that were successively shorter and shorter.

A similar story could be told about most human cultural advances. Transportation: from walking and running, to riding a horse, to railroad trains and steamships, to automobiles and trucks, to airplanes, to supersonic jets, with the last advances all happening during the twentieth century alone. Communication: from the invention of language, to writing, to the printing press, to Morse code, to radio, to television, to personal telecommunications—again, with nearly as much change since the year 1900 as during the preceding 1900 years. Food technology: from hunting, scavenging, and foraging, to simple agriculture, to the appearance of cities made possible by agricultural surplus, to irrigation, to mechanized and chemically assisted agribusiness. The list goes on, including similar patterns in the technology of warfare and killing, and ironically its mirror-image technology in medicine and life-preservation; the complexity of human social organization; independence from outside weather; as well as the growth of human population itself. Most of the time, human beings have managed to keep their balance and their poise, at least partly because even changes that appear rapid when viewed from a distance, often seem slow compared to the pressing moment by moment events of our daily lives. But in fact, the pace of the human cultural experience has been extraordinary, measured by any other standards.

If the time that human beings have experienced in truly rapid cultural change appears as a mere surface gloss upon the deep underlying substructure of millions of years of biological evolution, this is not to devalue its importance. Rather, quite the opposite: despite the fact that we have spent only a few moments as cultural creatures after a long previous lifetime as biological beasts, it is also a fact that we are now entirely enmeshed in that "gloss." It is also the point. In an instant of brilliant illumination, the lightning bolt of cultural innovation has suddenly transfixed and drastically changed our landscape. It may have been brief, but it has also been potent; indeed, its brevity highlights its intensity. For while Homo sapiens is a slow-moving, painfully evolving, tortoiselike Darwinian creature, the speed of Lamarckian cultural evolution is almost like the speed of light. If the interval between the invention of writing and the computer (perhaps 7000 years) seems long, it is nearly undetectable when measured against the interval between the "invention" of jaws from the old vertebrate branchial arches and the "invention" of binocular vision by our arboreal ancestors (perhaps 700 million years); and yet, the former may be no less important than the latter.

In exploring the paradox of Homo sapiens, we shall therefore be exploring the strangeness of time itself. Time is a duration of something, or perhaps an interval between things. When nothing is happening, we say that time stands still; when a lot is going on, time seems to speed up. A second is approximately the space between heartbeats. A year is about what it takes for a dog to grow up. When we say that it takes nine months for a person to develop from fertilized egg to newborn child, we mean that it takes less than one rotation of the earth about the sun, about 270 rotations of the earth about its own axis (causing alternating periods of light and dark on its surface), and perhaps 23 million times longer than it takes to blow your nose.

Einstein has shown that time itself appears to slow down for objects—or, in theory, people—traveling very fast relative to others who are, comparatively speaking, standing still. But for all of our history, all the time, there is no reason to think that the nature of time itself has changed; what has changed is the amount of goings-on that has occupied a given stretch of time, comparing say 50,000 years ago, and today. This is not to say that day to day events didn't happen then, or that lives weren't full. Rather, the occurrences of one life did not carry through substantially into succeeding lives. Experiences didn't accumulate. Each generation had to learn things pretty much anew, and pretty much the same things each time: how to make a kidney, a fire, a digging stick.

"The whole succession of mankind," wrote Pascal, "in the course of so many centuries must be considered as one selfsame man who exists always and learns continuously." In existing "always" (perhaps 50,000 years) Homo sapiens has indeed been learning continuously, and as we have seen, the amount that we have learned has been increasing exponentially. And during that time, we have essentially remained "one selfsame man" and woman, at least in our basic biology.

Our experience on earth, while we have persisted as one, selfsame species, can been divided into intervals of human lifetimes. Taking the biblical threescore and ten (seventy) years as our yardstick, the human experience has lasted roughly 700 lifetimes. Of these, our first 500 were spent without agriculture, without domesticated animals, without metals, living in caves or huddled under trees. Cities appeared only in the last thirty lifetimes, and for the great majority of the world's people, only the last two or three. Only in the last ten such intervals has one lifetime been able to transmit its wisdom to another, nonoverlapping one, by writing. Electricity and steam power were effectively harnessed only two lifetimes ago, about the same time as the Industrial Revolution. And automobiles, computers, jet planes, nuclear weapons, air conditioning, plastics, antibiotics, washing machines, department stores, the Superbowl, world wars, pesticides, assembly lines, and widespread literacy—experiences of a lifetime, all—are uniquely the cultural furniture of just our most recent one, the 700th lifetime.

We evolved rapidly as a species, squeezing more into our 700 lifetimes as Homo sapiens than in thousands that had come before, and more than any other species in the course of its existence. But even then, people of the past could be forgiven for not noticing that much had changed during their lifetimes. Because probably not much had. Now, it is impossible to miss it.

The zebra's leg-bone really hasn't been up in the air very long. An ape-man threw it up, and when it comes down, it is still an ape-man who must catch it. The hare and the tortoise, as different as can be, are nonetheless the same creature.

The interaction between human culture and biology is not simple or easily described. Culture supports biology, in such obvious ways as the encouragement of physical health, production of food, or of shelter. In other, more subtle ways, culture seems to mimic biology. As anthropologist William Durham has emphasized, human cultural patterns may incline toward activities that are fitness enhancing, even though the participants may not consciously be aware of the connection. It is also possible, as physicist Charles Lumsden and sociobiologist Edward O. Wilson have argued mathematically, that our genes hold culture on a leash, so that even our complex cultural and social arrangements are somehow restrained, ultimately, by our DNA. (If so, however, it should be obvious that the genetic leash is very long.) In addition, culture may occasionally go its own stubborn way, following patterns of development and elaboration that are responsive only to itself, and to certain rules of learning and cultural inheritance, as geneticists M. Feldmann and L. Cavalli-Sforza have argued.

Human culture and human biology are both so vast and multifaceted that they are bound to intersect in many ways, in various different points and along oddly projecting edges and planes. Accordingly, it seems naive to expect that any single theory will account for the nature of that contact. Given the peculiar in-between-ness of our species, it even seems naive to expect that we can ever be complete enough to know anything for sure, about either ourselves or the world around us. "Cultural man proposes," writes anthropologist Weston La Barre in his book *The Ghost Dance*,

> but reality disposes, for man is only another kind of animal. In a universe where the very stars whirl and wheel in places we discern only light-years later, and even the great planets wander exquisitively responsive to bodies we incompletely descry in the night, a necessary wisdom is humility, a knowledge that we do not know but can only say that, blessed and burdened with our current projective fantasies, this is the way it seems, now.

One generalization that seems safe, however, is that given the radically different rates at which they develop and the dramatically different processes which they embody, culture and biology—although forced together by their conjunction in the same organism, Homo sapiens—must only rarely be in perfect harmony. Or at least, to echo La Barre, this is the way it seems, now.

Human societies, throughout history and continuing into the present, have eagerly seized on many biologically influenced traits and extended them, nonbiologically, into the cultural sphere. If males are more likely to fight than females, then cultures are likely to dictate that men are *expected* to fight more than women. If people are biologically inclined toward nepotism (treating relatives more favorably than nonrelatives), then culture is likely to reify this inclination, too, prescribing detailed patterns of appropriate and inappropriate behavior. In short, despite the remarkable flexibility that is its hallmark, culture can also lead to hardening of the categories.

Culturally demanded traits have often been exaggerated beyond anything produced by biological evolution, leading to an important, widespread phenomenon that we might call "cultural hyperextension." Rather than setting itself in opposition to innate human inclinations, culture often seeks to mimic and extend these inclinations, in the process outdoing nature itself and going too far. In some societies, accordingly, it is not only expected and required that males be aggressive, and more aggressive than women, but this difference is hyperextended beyond anything that natural selection would condone. Among the Plains Indians of North America, for example, a young man had to prove himself by going without food or water for days, seeking visions. An aspiring Masai warrior had to kill a lion. Among some of the New Guinea tribes, only a human head would do. Although sometimes women were also aggressive in these societies, it seems clear that any existing male-female differences were hyperextended by their cultures. In many cultures today, they still are.

Ornithologist S. Dillon Ripley has described "aggressive neglect" among certain birds, and a similar situation occurs among fish. In such cases, adults spend so much time defending their territories and squabbling with their neighbors that they do not adequately provision their young, keep the eggs warm or (in the fish example) free of fungus infections. As a result, their own offspring are less likely to survive. Aggressive neglect is almost certainly a pathological condition; it would be selected against, biologically. But pathologies can readily "evolve" in human cultural systems, especially when they represent seemingly adaptive hyperextensions of real situations and legitimate concerns. Today, one might argue that America has been subjected to a nuclear age example of aggressive neglect, the Caspar Weinberger disease. Defending one's offspring has long been adaptive; likewise for defending one's group. It has even been appropriate to sacrifice other possible benefits—even, sometimes, one's life—in the pursuit of security from competitors or predators. But it is only by a grossly maladaptive cultural hyperextension of such tendencies that the United States invests enormous quantities of its national treasure in a futile attempt to coerce its adversary, while in the process undermining its strength and the real security of its own children.

Efforts to "biologize" human behavior, whether successful or not, have typically been attempts to identify the biological underpinnings of what we do. By contrast, biologically based critiques of human behavior have sought to focus on presumed tendencies on the part of human culture to inhibit or to fall short of our biological inclinations. Culture then stands accused of being insufficiently biological, and the message, typically, is that if only we could structure our societies more in accord with our biology, then all would be well. But in searching for the likely causes of our malaise, discomfort, and danger, and in fact, just seeking to clarify the human situation, it seems likely that one culprit is cultural hyperextension, the hare's tendency to run for miles in a direction that the tortoise has just taken a single step. In short, sometimes the problem is not that culture is insufficiently biological, but rather that it is excessively so.

Given the likely disparity between culture and biology, it is surprising that earlier thinkers have so often been inclined to accept the way things *are* as the way nature intended them, the way they have always been, or should always be. In his *Politics*, for example, Aristotle wrote, "It is evident that the polis [city-state] belongs to the class of things that exist by nature, that man is by nature an animal intended to live in a polis." Not surprisingly, Aquinas said the same thing about the medieval, Holy Roman Empire, and to many in the present day, the nation-state appears to be self-evidently the highest and most biologically appropriate political organization for our species. A perspective on human biological as well as cultural evolution may help us break free from such limiting "chronocentrism," the notion that current times are the key to all time.

As opposed to "culture," the word "civilization" derives from the Latin "civis," which means citizen of a city; it typically refers to more elaborate, advanced, and technologic systems. The first civilizations therefore began even more recently than the first human cultures, and there is little doubt that the entire structure of civilization has developed during a time period so brief it is unlikely that we ourselves changed biologically during that interval. Perhaps our biological natures are so fluid, so flexible, so infinitely malleable and adaptable that we are able to coexist comfortably with any cultures that we create, and any civilizations as well. In short, perhaps we have essentially no genetically prescribed human nature; if so, then we literally are what our culture makes of us, and by definition, we couldn't be uncomfortable, except perhaps insofar as cultural systems impose their own unique and distressing pain or anxiety, or subject us to the din of inequity.

But on the other hand, if there is a human nature, a Darwinian tortoise no matter how diffuse underneath the cultural trappings of modern human beings, then we can be virtually guaranteed that it coexists uneasily, at best, with our Lamarckian hare. The hare will almost certainly undershoot our biology, or hyperextend it. As we have already seen, there would be little if any difficulty exchanging a Cro-Magnon and a modern infant but great incongruity in making the same switch with adults of both cultures. This incongruity (literally: an inability to fit), between our biology, which has evolved by the laborious process of natural selection, and our culture, which has appeared with explosive speed through cultural evolution, is the root of nearly every human difficulty. It will be the focus of the remainder of this book.

CHAPTER FOUR

SEX

From Procreation to Recreation

The sexual life of the camel is deeper than anyone thinks.
In a moment of amorous passion, he tried to make love to the Sphinx.
But the Sphinx's posterior arrangement lies deep in the sands of the Nile,
Which accounts for the hump on the camel
And the Sphinx's inscrutable smile.

—Anonymous

Anyone watching monkeys in a zoo might think that they were oversexed. In fact, Lord Solly Zuckerman, a highly respected British biologist, once concluded that the social behavior of a troop of baboons was almost entirely determined by sex: frequent mountings by the rapacious males matched only by nearly constant solicitation from the nymphomaniac females. Sex seemed to be the social glue that held monkey life together.

Within recent decades, however, biologists, anthropologists, and psychologists have conducted numerous studies of the behavior of free-living primates in their natural habitats. This work has revealed sex to be a relatively unimportant part of their normal behavior. In fact, all female mammals have a short period of time, ovulation, when they produce one or more eggs that are capable of being fertilized and producing offspring. Among most animals, such as deer or birds, ovulation is limited to certain seasons of the year (spring in birds, autumn in deer). Among others—mice and men, for example—it continues throughout the year in a regular, predictable cycle. In either case, conception can occur only during these restricted times, separated by relatively long periods of infertility.

Females of most species are sexually receptive to the males only during this brief ovulation period and in fact, they may react quite aggressively to any sexual overtures outside this time. In almost all cases, however, the males are also selective in their advances, coming into season ("rut" in deer and elk) in synchrony with the time of ovulation ("heat") by the females. In addition, among most species the males are able to determine when the females are maximally receptive by being sensitive to chemicals produced by the females at this time. The extraordinary interest shown by male dogs in the rear ends of bitches in heat attests to this sensitivity, as does their interest in the urine of other animals.

Because sexual interest is generally limited to periods of fertility and animals can precisely identify these times, sexual intercourse is actually quite rare among animals in nature. This makes sense, because copulation involves a great expenditure of energy, and energy—the basic currency of life—is not expended without good reason. Furthermore, copulating animals are generally rather preoccupied with each other and are therefore more susceptible to being attacked by a predator. In many cases, when the animals involved are highly aggressive or predatory, there is even danger that the prospective lover will be attacked by its mate. So sex may be not only wasteful, but downright dangerous.

Among human beings, however, sexuality is unique in that despite its dangers and disadvantages, copulation is *not* limited to times when reproduction is likely. (In fact, under many circumstances it is limited precisely to times when it is *unlikely*.) This seeming anomaly is explained when we realize that sexual behavior in Homo sapiens has been liberated from the purely reproductive function that it serves in nearly every other animal. Just as George Bernard Shaw once said that youth is so wonderful it's a shame that it's wasted on the young, human beings have found sex too wonderful, or useful, to be wasted on reproduction alone. It has been modified to serve a "higher" function: maintaining and strengthening the bond between adults.

In Robinson Jeffers' magnificent, brutal epic poem "Roan Stallion," we are given the image of a woman who seeks something more gratifying than her disappointing relationship with an insensitive husband. She literally falls in love with a glorious stallion, and that relationship, although sexual, is also deeply suffused with religious awe and mystical transcendence. At the end, the stallion kills the woman's husband, and the woman, in a final recognition of her fidelity to the human race, kills the stallion. For her, killing the roan stallion was almost like killing God: as a human being, her sexuality had been modified to serve a higher function, not just a physical act, and certainly not just procreation, but the essence of relationship. The stallion, by contrast, is capable of no such leap, either emotionally or cognitively. Human beings, alone among animals, are.

To understand this discontinuity, let us go from the sublime to the anatomic, and consider the peculiar anatomy of our species. In assuming an upright posture we diverged radically from the traditional body plan of other mammals. Most mammals are supported by four limbs, with the internal body organs basically suspended like salamis on a butcher's rack from a long backbone that is normally parallel to the ground. But by getting up on our hind legs we forced a radical redesign of our skeleton, especially our pelvis, not only to accommodate a new attachment for the hip muscles, but also to provide a basinlike scoop supporting our internal organs. This modification restricted the space available for our birth canal, thus making childbirth in human beings a rather trying affair when contrasted with other mammals. The newborn's head is exceptionally large relative to the rest of its body and in fact, would probably be larger yet if not for the restrictions imposed by a narrow exit tunnel. This limitation of human brain size at birth has in turn necessitated a prolonged period of brain development following birth. (It also necessitates a lot of protein in the first few years of childhood. Protein deficiencies at this time can deprive the brain of needed development and result in permanent retardation. The low protein diets of many underdeveloped nations as well as the starvation in sub-Saharan Africa are a particularly serious problem for this reason, even for those who survive.)

The natural delay of much brain growth and maturation until after birth has further necessitated a prolonged period of infant dependence among human beings, during which time substantial adult attention is needed, not only to protect the helpless infants but also to provide necessary training and guidance. This is where sex comes in. What mechanism would keep the parental pair together, thus providing maximum security and training for the young? If sex were limited to ovulation only—as it is with free-living baboons today, or with the roan stallion's more traditional mates—it is unlikely that the family as a relatively stable male-female union would have evolved. Instead, our social system might resemble that of the modern baboon, with numerous males and females but no permanent association between them, only the transient union of male-female "consort pairs" when the females are ovulating and receptive. Or perhaps we would be emulating the wild horses, with a dominant stallion maintaining a harem of mares who respond sexually, but probably do not love.

Human beings are unique among animals in that we participate extensively in nonreproductive sex. In order to enhance its effectiveness in cementing a bond between adults, the physical act has been reinforced by emotional factors generally identified as love. Whether love can operate extra-sexually is a debatable point, since even Platonic relationships such as "love" of God, country, or parent may involve sublimation of sexual impulses.

We are also unusual, if not unique, in the degree to which ovulation is hidden. Even casual zoo visitors can determine the estrus state of a female monkey, typically by the enormously enlarged and brightly colored perineal region. Women not only lack anything comparable, but in most cases, the exact time of ovulation is even concealed altogether; delicate thermometers or chemical tests are in fact required to detect what in other animals is usually immediately apparent. Sociobiologist Nancy Burley has suggested a novel explanation for this paradox, based on a confluence of evolution and human consciousness. Thus, she suggests that early in human evolution, the connection between ovulation and pregnancy may well have been recognized, along with the connection between childbirth and pain, and also mortality. This latter awareness, among women, could have inclined them to *avoid* pregnancy by refraining from sex when they were ovulating. Now, here comes the interesting twist: women who were conscious of their own ovulation would have been selected against, whereas those whose ovulation was concealed would have left more offspring—unintentionally, to be sure—thereby selecting for what they did not know. According to this hypothesis, then, it is not so much biology and culture that are in conflict here, but biology and consciousness; and in Burley's scheme at least, biology won.

It seems likely that fear of pregnancy inhibited female sexual behavior, once the connection between sexual intercourse and reproduction was understood. It is also a truism that with the invention of cheap and available contraceptives, sexual behavior has been more liberated from its biological consequences than ever before.

Evidence for the unique role of sex in human life comes from study of its physical components. Ejaculation by the male involves the expulsion of sperm in a rich, nutrient liquid suspension (semen), providing the microscopic sperm with food, energy, and a fluid medium through which they might swim to their ultimate rendezvous with a ripe egg within the female's body. Vaginal fluid is acidic, potentially lethal to sperm; male ejaculate therefore compensates by being appropriately alkaline. Discharge of the necessary fluid requires vigorous muscular contractions by the tubes that carry it, and the pleasurable sensation accompanying its release is closely connected with the release of tension as the tubes are physically emptied. It is thus interesting that the physiology of male orgasm does not differ significantly from humans to most lower mammals. The "need" for sexual release and satisfaction upon its accomplishment can be explained in evolutionary terms as due to the selective advantage of sending one's reproductive products on their way. (Individuals who copulate successfully are obviously more likely to leave offspring.)

The female version is a different story. There is no compelling evidence for a female orgasm in any animal other than Homo sapiens, although recent data suggests it as a possibility in certain monkey species and in the dwarf chimpanzees. It seems that in this regard we are again unique, or nearly so. By a great diversity of mechanisms, all initiated and then strengthened by the evolutionary force of natural selection, females of every species are variously motivated to copulate at the appropriate times. But in nearly all cases, there is no indication that they ever "get" anything out of it, other than simply to satisfy the motivating factors that led them to copulate in the first place. And ultimately, of course, getting pregnant as a result.

Among human beings, however, women are capable of experiencing orgasms of their own, although quite different from male ejaculation. Skinnerian psychologists might describe this phenomenon as "positive reinforcement," in that once experienced, it tends to increase the probability that the individual will repeat the behavior immediately preceding it. Orgasm, then, gives women a direct stake in copulation, making regular sexual activity more likely and with it, the increased coordination of male-female behavior that is the evolutionary payoff.

It is also worth noting that female orgasm is more difficult to achieve than its male counterpart. Thus, whereas men can often reach orgasm in a few minutes or even less, women typically take longer. Among certain animals it appears that dominant males often take longer to ejaculate than do subordinates. The mounting of a female grizzly bear by a dominant male is rather a leisurely affair, whereas subordinate males, obviously hurried because they are harried, spend most of their mating time looking over their shoulder, alert for the possible appearance of the dominants. Not surprisingly, they ejaculate very quickly, before they could be interrupted. Female grizzlies, to our knowledge, do not have orgasms, but it seems likely that even if they did, the sexual style of subordinate males would be unlikely to elicit very much response.

Assume now that sex among human beings serves social bonding as well as reproduction—recreation as well as procreation. Then it is not surprising that female orgasm, which may well be related to bonding, is more likely when the sex act itself is prolonged, and hence, when the male partner is more likely to be successful, effective, and desirable, the human equivalent of social dominance among animals. It remains to be seen, of course, whether "latency to ejaculation"—the time between penetration and ejaculation—in men is related to experience, self-esteem, and something analogous to social dominance. But it may be significant that premature ejaculation is particularly a complaint of young, inexperienced males, that ejaculation takes longer among older males, and that sexual responsiveness among women is much greater after lengthy sex—characteristic of dominant male animals—rather than rushed and harried sex, more typical of younger subordinates.

Many human physical characteristics attest to the importance of sex and indicate our potential for a high level of sexual fulfillment. Thus, all mammals feed their offspring milk, made from their mammary glands. But only human beings have breasts. Among all other mammals, the milk-making glands are trivial and inconspicuous except when the females are actively lactating. By contrast, human beings are unique in possessing well-developed, protuberant breasts even among nonreproductive individuals. There seems little doubt that this development is related to our exaggerated sexuality and to our elaboration of nonreproductive sex as well. By exaggerating a trait that is biologically associated with successful reproduction, natural selection may well signal reproductive competence; otherwise, there is no biological reason for female breasts to become enlarged at the onset of menstruation; they could just as well wait until pregnancy and lactation, as in all other mammals. But in human beings, sexuality has developed alluring aspects, quite independent from reproduction.

One needn't be a trained evolutionary biologist to speculate in this manner. Stephen Dedalus, the youthful hero of James Joyce's *A Portrait of the Artist as a Young Man*, suggests the following hypothesis while musing with his friends on the nature of female beauty (as perceived by men):

> [E]very physical quality admired by men in women is in direct connection with the manifold functions of women for the propagation of the species [modern biologists would here substitute "propagation of their own genes"]. It may be so. The world, it seems, is drearier than even you, Lynch, imagined. For my part I dislike that way out. It leads to eugenics rather than to esthetic. It leads you out of the maze into a new gaudy lectureroom where MacCann, with one hand on *The Origin of Species* and the other hand on the New Testament, tells you that you admired the great flanks of Venus because you felt that she would bear you burly offspring and admired her great breasts because you felt that she would give good milk to her children and yours.

Dreary or not, there may well be a good deal of truth in such notions.

It is noteworthy, however, that prelactational breast size among human beings actually does not correlate with milk production, so that the "great breasts" of Venus to which Stephen Dedalus refers may be potentially false and misleading advertisements. The nonlactating female breast is composed almost entirely of fat, whereas it is glandular tissue—which develops during pregnancy—that produces the milk. Large, fatty breasts may therefore be biologically deceitful, giving the illusion of mammalian plenty, but when push comes to shove, not necessarily any more substance.

Just as women have the largest breasts of any mammal, men have the longest penis of all primates, a characteristic that is as distinctive as our large brain, but to some, less elevating. Maybe the oversized penis of Homo sapiens is an adaptation to deposit sperm deeply into the female reproductive tract, necessitated by the tilting of the female pelvis that resulted from walking upright. Or perhaps there is some other reason; in any event, men seem to worry about their penis size about as much as women worry about their breast size. (Ironically, such concern on the part of men is also about as irrelevant, biologically, as the breast obsession of Western societies: once erect, there is very little difference in penis dimensions.) Regardless of our biology per se, it is clear that we have achieved substantial cultural hyperextension in our sexually related phobias and preoccupations.

As if all this hyperdevelopment of breasts and penises was not enough, our hairless body, covered with a great number of sensitive touch receptors, provides additional opportunity for enhanced sensual input. And the great dexterity of our fingers and lips (again, surpassing that of all other animals) also enables delicate and varied stimulation of the partner. We are clearly the sexiest animals on earth, by virtue of our biology alone.

Nonreproductive sex is an exceptional human attribute and evolution has obviously conspired to provide us with the opportunity and motivation for such activities, emancipated from reproduction. It is therefore ironic that the Catholic Church, for example, specifically considers such behavior to be immoral because it is "bestial" and "catering to our animal instincts." *Animals* are indeed limited by their instincts. They engage in sex only when it serves reproduction. People, by contrast, are in fact being uniquely human when they engage in sex for its own sake, for the higher relationship it helps achieve between a man and a woman. Sex limited to reproduction, on the other hand, is the norm for animals but not for people.

This is perhaps the most glaring example of conflict between the biological and cultural sides of sex. There are others, however. No biological phenomenon, in fact, has been more heavily overlain by cultural embroidery.

It is impossible to say which of the many marital arrangements practiced by different human cultures is the most biologically "natural." It may well be that local diversity in such arrangements is the biologically "correct" situation. Among monkeys and apes, different species practice different social systems and often, the same species may differ from place to place. In such cases, there seems to be adaptive significance to each particular arrangement. Living things have been selected to engage in different social systems depending on which ones are most successful in any given ecological situation. Yellow baboons forage through the dry east African savannah in large social groups with the males providing maximum protection for the females and young. By contrast, mandrill baboons of moist, forested west Africa travel in small nuclear family groups that appear to be monogamous. Among human beings, social and marital arrangements are determined by cultural and especially religious convention, without obvious regard to the regionally optimum ecological arrangement. Thus, the Moslem's maximum of four wives on the one hand, the Judeo-Christian policy of only one, on the other, may or may not be biologically sound, but both were probably arrived at for reasons other than their ecological and evolutionary utility.

The "double standard," allowing greater liberty for sexual experimentation and dalliance to men than to women, is well established in most human cultures. In general, men are more easily stimulated than women, and are the sexual aggressors. Rape is characteristically done by males to females, not vice versa, and this is not simply a result of the former's greater physical strength. Rather, it reflects a basic evolutionary consequence of maleness versus femaleness. Similarly, it is not just coincidental that prostitutes are usually females, visited by males—only rarely vice versa—

and that pornographic magazines are regularly purchased by millions of men but have much less market among women. The cultural differences, with all their complex embroidery, are almost certainly overlain on some clear-cut biological differences as well.

Females of every species produce eggs, each of which may be thousands of times larger than the male's sperm. Among such animals as fish, amphibians, reptiles, and birds this contrast in size is especially striking, since the egg is easily visible to the naked eye, whereas all sperm are microscopic. Among mammals the egg is considerably smaller than in lower forms, since the developing young are nourished from the mother's blood and the egg need not contain great food reserves. Nonetheless, the nutrient-laden egg dwarfs the tiny sperm. The production of eggs in all animals, including ourselves, clearly involves a much greater investment of metabolic energy than does the manufacture of sperm, especially since once fertilized, an egg must receive massive subsequent investments from the mother's body. Some birds may produce eggs totaling fully one third of their own body weight. An adult woman may produce about 400 eggs during her lifetime, whereas a man will discharge 100 to 300 *million* viable sperm at a single ejaculation. Because of their small size, the latter can be produced in fantastic numbers. In addition—and perhaps, most important—responsibility for both pre- and postnatal care of young almost always falls upon the female. Among birds, prenatal care involves the production of large, well-supplied eggs, whereas among mammals, the female nourishes developing young from her own bloodstream and then later, by the milk she produces.

Seen from any angle, in most species the female has a greater stake in the outcome of any one copulation than does the male. She has invested more metabolic energy in egg making and she produces far fewer eggs. Hence, she stands a greater chance of losing out entirely if she is unsuccessful in securing proper fertilization for them. She literally has all her eggs in a limited number of baskets and—just as literally—she must bear the responsibility of a mistake. By contrast, the male's biology gives a degree of sexual license. It is no accident, then, that males are characteristically more easily aroused and more prone to be sexually forward than are females. Thus, even though male sexuality is typically cued to female receptivity, during the breeding season males are normally less discriminating than females.

Certain species of orchids get themselves pollinated by producing flowers that mimic female wasps. The sexually aroused male wasps try to copulate with them, get dusted with pollen, then proceed to the next acquiescent and deceptive flower, whereupon they unknowingly transfer the pollen between their successive floral "lovers." It is noteworthy that we don't find flowers masquerading as male wasps, simply because no self-respecting female would be so easily misled into copulating with a flower; only with another wasp. After all, she has large and precious eggs at stake; he has only some cheap and easily replaced sperm. If those creatures he finds so alluring were female wasps instead of flowers, he would be spreading his genes, and at little cost to himself. As it is, the wasp's maleness, with its eagerness and lack of discrimination, is readily exploited by the flowers.

In many other species, males will court almost anything, whereas the females are more demure and fussy about their choice of mates. Male crabeater seals, for example, congregate around females and their unweaned pups, perched on ice floes. The males try to mate; the females resist. The males bite the females; the females bite back. Both may get covered with blood, as the male seeks to induce the female to accept his sperm, while the female resists, presumably because it is not in her evolutionary interest to get pregnant too soon, since that would require weaning her pup before it is adequately grown. The course of true love among animals doesn't always run smoothly. Animal courtship often involves breaking down the resistance of the female, with stimulation of the male being relatively unimportant, because it is so easily achieved. Once again, it is easily achieved because as with the wanton wasps, the cost of male error is small and the potential benefit of success, large.

In at least one species of insect, the "Mormon cricket," the tables are turned: males are shy and females are sexually pushy. In this species the male transfers a large, gooey, nutrient-rich structure called a spermatophylax as part of mating; the spermatophylax represents a metabolic investment comparable to that of the female's eggs in many other species. Not surprisingly, therefore, in this case the male acts "femalelike" and vice versa. Female Mormon crickets must climb on top of the males before the latter will consent to mate with them, and the precious spermatophylax will only be transferred if the females in question are heavy enough; heavier females make more eggs, so in this unusual (but understandable) case, the male holds out for the mate with the biggest dowry.

Most of the time, however, it is the males who are eager to transfer their abundant sperm and the females who jealously guard access to their precious eggs.

The usual male-female difference in sexual style is also good biological evolutionary strategy. When males produce large numbers of sperm and can easily replace any that are discharged, natural selection would favor "playing fast and loose." By freely expending sperm on the slightest pretext, they would stand a greater chance of leaving offspring in the future.

In contrast, an error by a female may seriously handicap her reproductive future. Although it is to the male's advantage to maximize the dissemination of his sperm, good female strategy calls for optimization. Successful matings between males and females of different species produce offspring—hybrids—that are generally inferior to either parent in a variety of ways. As a result, the careless female who falls for the charms of a male from the wrong species has expended her precious genes (or, more to the point, her valuable reproductive investment) on a loser. Since offspring from such females are less likely to be successful, females with a genetic tendency to be less discriminating will leave fewer offspring. Eventually such types will decline in numbers, replaced by those of the more careful females who trusted their genes to male partners more likely to bring them success.

For example, domestic dogs are all members of the same species; hence they can exchange genes with one another. But artificial breeding programs have divided *Canis domesticus* into dozens of breeds, some of them dramatically different from each other. Take a female Pekinese or a tiny Chihuahua: in heat, she will readily accept any male dog, even a St. Bernard or Great Dane.* By accepting such an oversized mate, the miniaturized canine mother will have almost certainly signed her death warrant: her developing offspring will be too large to pass through her birth canal, and both bitch and litter will likely die unless someone performs a cesarean.

This is clearly a very artificial situation. Under natural conditions, a single species such as the domestic dog would not have differentiated into such very different shapes and sizes. Under current conditions, the small bitch who allows herself to be inseminated by a giant male is an evolutionary loser; such lack of discrimination would be strongly selected against. For her paramour, however, the situation would not be tragic at all. Dogs do not form long-lasting pair bonds, so if his mate dies in childbirth, the undiscriminating Lothario has not lost very much. This may seem overly harsh judgment, and certainly unfair in its punishment of the female while the equally "guilty" male gets off scot free. But evolution is not concerned with adjudicating morality, but rather with differential reproductive success.

We would expect therefore that under natural conditions, selection would tend to produce choosy females and relatively undiscriminating males. In many cases, a female whose mating preferences are overly lax may produce offspring that fail to develop normally and are aborted. Or the offspring may be perfectly healthy, but sterile, such as a mule which—true to its reputation—stubbornly refuses to pass on its parents' genes.

*In such cases, the physical mismatch in size between penis and vagina may make mating very difficult, however.

THE PRECEDING is not intended as a justification for the double standard in human society, or as a plea for its further elaboration. But it may help to explain an otherwise puzzling basic difference in the sexual behavior of men and women, even acknowledging that the human situation is much more complicated than it is among romantically inclined French poodles. Opportunities for disadvantageous hybridization may have existed among our Australopithecine ancestors, and even Cro-Magnons may have had the potential to interbreed with the Neanderthals. In fact, fossil skeletons at Mt. Carmel, in modern Israel, suggest that this may have occurred. But modern Homo sapiens is not seriously subject to any such temptations. Nature's scarlet letter, meted out automatically to errant animals, is handled by cultural means in the human species.

What does all this have to do with the conflict between biology and culture among Homo sapiens? As we have seen, natural selection, working through the advantage gained by a strong bond between parents, evolved mechanisms to achieve and maintain that bond, most notably the liberation of sex from its purely reproductive function. This may well have been achieved in part by endowing women with the capacity for orgasm, as well as the attendant emotional attachment of both partners. But by enhancing the female's sexual responsiveness, evolution has made the female receptive to males other than just her chosen mate. Thus, although it is characteristic of most animals (including humans) for the male to be more easily excitable, seduceable, and hence, likely to philander, the phenomenon of intentional female philandering is somewhat unusual in nature, but conspicuously well developed in human beings. Ironically, while selecting for female responsiveness as a means of maintaining the pair bond, evolution may also have set in motion forces that among many human cultures have traditionally acted to weaken that pair bond.

Different societies respond to a woman's sexual infidelity in almost every way imaginable, ranging from the death sentence to indifference to (very rarely) approval. Philandering by the male elicits equally varied responses, although among most societies it is generally regarded as less serious than comparable female behavior. This may well be due to the fact that male promiscuity has a firmer biological basis. The woman who gets pregnant has a substantial reminder of her behavior, and men whose mates get pregnant have lost some of their own potential fitness. The outrage of the cuckolded husband, often abetted and even encouraged by societal norms that may condone brutality or even murder as a response, represent a heavy-handed dose of cultural hyperextension.

Evolution, however, has set a potential trap for the dallying man as well. Among nonhuman animals there is little or no evidence of emotional concomitants to sexual behavior by the male, excepting the physical urgency associated with orgasm itself. But selection for the pair bond in human beings has also had subtle effects on the man as well as the woman, endowing him with a capacity for strong emotional attachments that may well equal hers. Clearly, such attachments are not universally felt, either by all human cultures or by all individuals within any given culture, or even by a single individual in all cases. But the "eternal triangle" is a fairly common occurrence, in which a man may love two women; or a woman, two men. Biological evolution may contribute to attachments that are often unacceptable to the standards of cultural evolution.

Prior to Western colonialism and Judeo-Christian social imperialism, the vast majority of human societies were in fact polygynous: the preferred marital system involved one successful man mated to several women. Given the biology of maleness and femaleness, it is possible for such arrangements to enhance the evolutionary fitness of all—except for the excluded bachelors. On the other hand, polyandry is predictably rare: there are very few societies in which one woman is typically mated to several men. Such an arrangement would likely reduce the fitness of the participating males, since it would lower the probability of any single one of them being a father; at the same time, it would do little for the fitness of the woman, since her maximum reproductive success is unlikely to be greatly enhanced. She can only have one child at a time, no matter how many husbands.

Human biology, moreover, is entirely consistent with the suggestion that Homo sapiens is a mildly polygynous species: men are somewhat larger and constitutionally more aggressive than women, and women become sexually mature earlier than men. This pattern is widely found among polygynous species, in which males are selected to delay engaging in vigorous male-male competition until they are old and large enough to have some chance of success. Bull elk, for example, do not compete for a harem of females until they are several years older than the females with whom they will eventually mate.

In this regard, it may appear that culture is the villain, often unreasonably frustrating the natural, healthy, biologically evolved proclivities of our species. But what is "natural" is not ipso facto good. As human beings, we have the right, perhaps the obligation, to choose monogamy over polygyny, or a uniform sexual standard over the double one. Many things that are biological, and hence, very "organic," are not necessarily very pleasant: typhoid, forest fires, gangrene, and hookworms, for example. If the disharmony between biology and culture is ultimately responsible for most human dilemmas, this does not imply that biology should therefore be permitted to win. If Homo sapiens is to win, however, then it might help to become more truly "sapient" about this fundamental conflict.

The biological connection between sex and love is almost certainly a result of the fact that the initial "goal" of evolution was to maintain and strengthen the pair bond, for the ultimate benefit of the children and hence, the fitness of the parents. Insofar as cultural restrictions serve to discourage adulterous or promiscuous behavior, thereby preventing erosion of the parental pair bond, such restrictions may ultimately be consistent with the thrust of biology as well. Furthermore, it may well be that excessive sexual freedom is ultimately doomed to impair the emotional integrity of its participants . . . not because it defies certain fundamental ethical or religious (that is, cultural) precepts, but rather because it ignores the basic connection between sex and love promoted by biological evolution.

Given that natural selection has succeeded in adding a potent emotional and pair-bonding function to the straightforward reproductive role of sex, it would be foolhardy to expect the two functions to be easily separated. Human beings achieve enormous personal, emotional benefits from sexual behavior, but these benefits can become hindrances to "pure, natural, free-love" when emotional ties are not only unsought but also unwanted. When culture prescribes sex without such ties, physical relationships can usually be made just that: physical, and devoid of personal bonding. But the ultimate cost may be great in terms of decreased ability to establish warm, loving relationships in the future. The classic case of sex without love is of course the prostitute, who by severing the biological connection and concentrating on the former, may suffer diminished capacity for the latter as well.

Biological evolution undoubtedly has had a profound impact upon human sexual behavior, selecting for its pair-bonding functions in addition to its purely reproductive effects. However, we must remember that biological evolution is a terribly slow process, easily outdistanced by cultural change. When natural selection began operating on the genetic systems of Homo sapiens and our precursors, fashioning our response to sexual encounters, human culture was very rudimentary, almost nonexistent. To primitive human beings, with virtually no complex society to surround and nourish us, the maintenance of a strong male-female pair was of crucial importance to the survival and success of children. Since that time, our culture has evolved so much that many of our biological characteristics may be anachronisms, as out of place as the bones of *Tyrannosaurus rex* in midtown Manhattan. But whereas those bones reside in the American Museum of Natural

History, some of our current biologic fossils still lurk within ourselves.

Evolution's payoff is in successful reproduction and, in turn, the successful reproduction of one's offspring and other relatives. A fatherless prehistoric child was at a real disadvantage, unlikely to be well provided with food or protected from enemies. Unless immediately adopted, an orphan was as good as dead. But in modern society, a successful paternal bond is only indirectly related, if at all, to the ultimate reproductive success of its offspring. Unwed parenthood is common, and the wealthy movie star or welfare mother will both probably produce reasonably normal, healthy children, eventually capable of having children themselves. Many cultures admittedly continue to discriminate against bastard children, and their parents, while supporting institutions such as religion that encourage relatively permanent male-female bonding. Thus, cultural defense of marriage may have developed in large part in support of the important biological function of marriage, the production of successful offspring. But the cultural innovations that are rendering the biologically evolved bonding of parents outmoded may well be undercutting the need for these cultural controls as well.

The institution of marriage also developed in support of cultural practices, notably the oppression of women, which are themselves facilitated by male-female differences. But other cultural developments such as schools, personal wealth, and government assistance, are drastically reducing the importance of the preexisting biological function. As a biologically necessary phenomenon, the male-female bond may be increasingly outmoded, even as we retain a deep-seated tendency nonetheless to respond strongly to real and potential partners. Parental bonding, in short, presumably began as a means to an evolutionary end, and it quickly incorporated sex as a useful tactic. Now, the end—successful reproduction—may well be achievable without the means, but we nonetheless pursue social and sexual bonds at least as much as reproduction itself.

Human beings still possess a coccyx, or tail-bone, as a remnant of our mammalian ancestry. In modern Homo sapiens, this structure has decreased since it is no longer selectively advantageous. Similarly, the appendix and tonsils are probably rudimentary compared to a time in our evolutionary past when they conferred some distinct advantage. As biological characteristics cease to be of value, they cease to be included in the genetic constitution of living things. Essentially, this is the Second Law of Thermodynamics at work: any complex structure, whether coccyx or the federal Constitution, requires an input of energy to keep it operating. Without natural selection or human consciousness providing regular input, nonrandom structures tend to decay: as the physicists put it, their entropy increases. But this process takes time, and living things, no less than governments, carry a lot of deadwood around with them. We might anticipate that as a result of present cultural trends emancipating a child's biological success from the success of its parents' social relationship, selection for biological factors maintaining that relationship would eventually decline. Biological evolution for the male-female bond may in fact be about to reverse itself. Clearly, its adaptive advantage (the production of successful children) is less important now than during our evolutionary past.

The precise role of biology in this entire business is difficult if not impossible to assess. Although impatient youth may bewail the conservatism of cultural change, our institutions are more likely to respond promptly to such drastically altered facts of life than is our genetic makeup. How they do, and indeed whether they do, is much more up to us than to our DNA. Ultimate success of male-female pair bonding in Homo sapiens may well depend upon the direction of cultural evolution and upon whether this bonding will be seen as worth maintaining for its own peculiar values. At one time, not only did the ends justify the means, they produced the means; now, the means may well have to stand, if they are to stand at all, as ends in themselves.

It is ironic that even as biological evolution may be inclined to eliminate the male-female bond because of its newly diminished utility, early cultural dissolution of it for the same reason may be bedeviled by the biological factors which are still tuned to maintaining the bond. It is certainly a very complicated story. Human culture has been exposing the adult pair bond to an increasing array of stresses. For example, we have already seen that if we did not differ significantly from other animals we might expect nothing from sex but procreation. But our evolution has provided other possibilities and accordingly, sexual behavior is now expected automatically to produce profound, "meaningful" relationships, experiences that are enriching almost to the point of mystical significance. We are in many ways a uniquely self-conscious animal, acting and simultaneously aware of ourselves as we do so. When it comes to experience as intense as sexuality, it is thus not surprising that we appraise performance, both our own and our partner's.

Such evaluative self-awareness may constructively improve any situation. But applied to sex, for example, it may also create temporary impotence, frigidity, and/or chronic dissatisfaction, the feeling that one is somehow missing something. Ironically, the proliferation of literate, scientific sex manuals has probably exacerbated the situation by heightening public awareness of what is "normal," and what, and how often, they should expect from their partner, themselves, and their experiences.

Another significant strain is imposed by the culturally evolved frequency of human encounters and interactions. Our genetic systems undoubtedly evolved through hundreds of thousands if not millions of years during which normal social interactions were limited to the other members of our primitive hunting and gathering bands. These probably contained no more than several dozen members each. Individuals within each band were certainly well known to each other. Even the occasional interaction with members of a different social unit probably constituted either a battle or a reunion of old acquaintances, since for better or worse, groups probably maintained a regular orientation relative to each other.

Thus, while the choice of sex partners may have been limited, so was temptation.

Today's human being may personally encounter literally hundreds, even thousands, of other people every day, many of whom may be quite attractive. This is something quite new to our experience as a species, although it is less obvious than computers, nuclear weapons, or nation-states. But whether identified or not, our genetic system almost certainly has not had time to evolve biological defenses against such hyperstimulation. Modern advertising barrages us with the images of our most alluring specimens in constant effort to generate the desired longing, envy, expectations, associations or simply to attract our attention. And given the effect of novelty on sexual behavior, attract our attention it does.

For nearly all species, sexual performance declines upon repeated experience with the same partner; however, among animals as diverse as rats and cattle, it can be returned to high levels upon exposure to a new one. It is said that when Calvin Coolidge and wife were separately touring a model farm, Mrs. Coolidge was impressed with the frequency with which the rooster mated: "Please point that out to Mr. Coolidge," she suggested. The President, in turn, asked whether the rooster was always mating the same hen. No, he was told, many different ones. "Please point *that* out to Mrs. Coolidge," he ostensibly replied. Students of animal behavior now speak of the "Coolidge effect," whereby sexual behavior, especially of males, is increased with a new partner. We obviously cannot say whether animals feel "bored" with a well-known mate, but they are clearly invigorated by a new one. To some extent at least, a similar phenomenon may occur among human beings. The classic "seven-year itch" reflects a time in marriage when other stimuli may be sought, and the marriage of old (usually wealthy or prominent) men to much younger women is often accompanied—if not precipitated—by the increased sexual vigor that usually ensues . . . if only for a short time. There is little known about the effect of novelty on sexual behavior in female animals, or on women either. It is likely to be similar to the male, although perhaps somewhat less intense. This is because, returning once again to the biology of maleness and femaleness, males (of most species) can increase their biological success by mating with a variety of partners and being relatively indiscriminate. A copulation with someone new may result in pregnancy, to the male's advantage, especially if he can avoid providing any subsequent resources or assistance. We are therefore biologically susceptible to, and evolutionarily unprepared for, such new stimuli, while at the same time our culture bombards us with them. It is not surprising that divorce rates are highest among the subculture most subjected to such disruptive sexual stimuli, the movie-star community.

Students of animal behavior have long recognized that many species possess a genetically determined susceptibility to particular stimuli that are often provided by another member of the species. Upon appearance of the correct signals, the other animal responds automatically with some basically instinctive behavior. The stimuli that elicit this behavior are referred to as "releasers" because they appear to release behavior that is fully developed within the animal, just waiting for the proper combination to let it out. In the model proposed by the famous ethologist Konrad Lorenz, a releasing stimulus permits a particular behavior to flow just like pulling the chain releases water accumulated above an old-fashioned toilet bowl.

For example, male European robins will attack a tuft of red feathers mounted on a stick and placed inside their territory, while ignoring a much more realistic stuffed robin that lacks the color red, which apparently releases robin aggressiveness. Crows will sometimes attack someone carrying a bit of black cloth, apparently because the unwitting victim has stimulated the "crow in distress" releasing mechanism normally reserved for predators such as hawks or owls. These reactions are automatic, not rational. This is why they can be produced by simple models with only the vaguest resemblance to a live animal, so long as the appropriate releaser is present. Similarly, among the North American woodpeckers known as yellow-shafted flickers, the males and females look almost identical except that the male sports a black streak on each side of his face, looking for all the world like a mustache. If an equivalent streak is painted onto a captured female, she will be attacked viciously by her mate. Possessing the releaser that says "male," she stimulates automatic aggressive behavior from her own consort.

Biologists have also noted that if the appropriate characteristics that make up a releaser are artificially exaggerated by an experimenter, animals will often prefer them to the naturally occurring signal, or will perform their particular behavior more intensely or for a longer time. These exceptionally successful manmade signals are called "supernormal" releasers. Many nesting birds, for example, will prefer to sit on the biggest egg possible, ignoring their own in favor of a larger experimental model. The oystercatcher, a common robin-sized shorebird, will apparently forget her own eggs and perch ridiculously on top of a huge artificial egg the size of a watermelon.

Another example: among birds that produce helpless young, such as robins and sparrows, parental housekeeping chores often include removing the shiny fecal sacks produced by the nestlings. If bird banders place the traditional shiny metal ring around a young bird's leg, the parent may literally attempt to throw the baby out with the bathwater, despite the protesting chirps from their own offspring.

As to our own species, if certain physical characteristics serve as releasers to human beings, their exaggeration—as for instance the 40-inch chest of the topless dancer or Playboy bunny—represents a culturally developed supernormal releaser. Our susceptibility to releasers is probably expressed in many other ways, going far beyond sexual stimuli. For example, Lorenz has pointed out that we tend to regard as "cute" and "cuddly" any picture of a child or animal with unusually large head and eyes, and an absence of protruding ears or nose: just look at any child's doll, or the cartoonist's representations of children. In such cases, the designer is playing upon our response to supernormal releasers by exaggerating those stimuli that normally release nurturant behavior in human beings.

Culturally produced supernormal releasers may provide enhanced pleasure to human beings who intelligently manipulate them. Ironically, however, they are also especially insidious precisely because they can be overridden. They lack the obvious automaticity—sometimes comic and absurd—of an incubating oystercatcher, a pugnacious robin, or a nest-cleaning sparrow. And therefore, we may well be susceptible to subliminal supernormal releasers, without realizing that they are acting upon us. If instead, like the engaging Mr. Toad of *The Wind in the Willows,* we became maniacally insane and uncontrollable when exposed to certain releasers (like that unhappy amphibian's "motor-cars"), then we could at least be on guard. But we are not directly at the mercy of our own mental mouse-trap response patterns. And as a result, paradoxically, we are almost defenseless against the subtle, hidden persuaders of our own making.

It might be argued that the evidence for genetically mediated susceptibility to releasers in human beings is thin indeed. Although female breasts have sexual significance in Western society, for example, a good case can be made that they are much less erotic in certain African societies where they are not provocatively hidden as a matter of course. It might therefore be best to modify the concept as it applies to humans, and to identify "cultural releasers"—stimuli that come to have specific behavioral significance largely as a result of cultural practices—with only possibly a faint hint of genetic underpinning. The particular stimuli would then vary from culture to culture.

Because of its peculiar quality of intentional directedness, distinguishing it from purely biological creations, human culture has a further potential: after generating culturally defined releasers, presumably by building on existing biological predispositions, societies can then proceed to exaggerate or hyperextend certain characteristics, thus eliciting a heightened response. The result? Another major source of potentially disruptive sexual stimuli, notably in the exaggeration of body features associated with sexuality. Thus, red lips have been emphasized with lipstick, eyes with eye shadow and mascara, and silicone implants can literally hyperextend the female chest. We are constantly subjected to these supernormal characteristics. Alternatively, we subject ourselves to these culturally generated characteristics *because* we are so susceptible to sexuality. In any event, not only do we produce cultural releasers, but we make them supernormal, and they both amuse and bedevil us.

> *In matters of sex, everything you can possibly imagine has occurred and much that you cannot imagine.*
>
> —ALFRED KINSEY

WE ARE among the few animals that show substantial variety in lovemaking; copulation among most animals is usually stereotyped and characteristic of each particular species, although some animals such as gorillas are quite inventive. The possibilities of copulation are clearly limited by the physical structure of our bodies and our genitals, but beyond that, human cultural ingenuity is almost boundless. Despite the great variety of lovemaking postures demonstrated by Homo sapiens, however, the ventral-ventral position appears to be the most common and generally popular. By contrast, the most common copulatory position among vertebrate animals involves the male climbing upon the female's back, the celebrated "doggy position." We are also unusual in employing the ventral-ventral position at all. In addition, we are unique, of course, in having such cultural "how to" books as the *Kamasutra* or *The Joy of Sex*. But if such sexual smorgasbords were part of our genetically mandated repertoire, we probably wouldn't need the books.

Among the many changes necessitated by our insistence on walking upright was elaboration of the gluteus maximus, the rear-end muscle. This may have hindered "normal" dorso-ventral copulation, possibly resulting in some ventral rotation of the vagina, thus facilitating face-to-face lovemaking. This basic postural change may also have been related to the very important connection between sex and emotional attachment, which was evolving at the same time. In short, face-to-face sex is personal sex. Shakespeare's "beast with two backs" is not really a beast at all, but rather, two human beings who are likely to know each other and who are in the process of getting to know each other better yet.

It is instructive to observe copulation (usually dorso-ventral) in the monkeys and apes: the female is strikingly uninvolved in the proceedings. If she happened to be eating when the male first mounted, she may continue to chew unconcernedly throughout the act. In fact, among the baboons and rhesus monkeys, females have been observed sexually soliciting an adult male who is eating something desirable. While the male is then occupied with his mounting, the simian Siren steals the food. With ventral-ventral copulation, such nonchalance is less likely and intercourse is more personal. The hands, for example, are available for fondling and manipulation of the partner, but for little else, while the eyes are facing each other, not gazing around for friends, enemies, food, or whatever. As ethologist Desmond Morris has emphasized, ventral copulation may thus be another genetically influenced mechanism for eliminating the emotional disinterest of animal matings and insuring the uniquely human sex/love relationship.

On the other hand, it is noteworthy that dorso-ventral copulation, even among human beings, is still the most likely to lead to conception, although it is definitely not the most popular. This may itself reflect the extent to which sex has been emancipated from its strictly reproductive function.

Not only are we unusual in making love belly-to-belly, but we also prefer privacy when we do it. In fact, this insistence is so ingrained that we may be surprised to learn that it also is unusual among animals. Copulating pairs of monkeys and apes do not appear to value their privacy or make a particular effort to hide their activities from the other group members, although among some species, consortships are formed during which male and female take a mini-honeymoon together, away from the others. The tendency for private and concealed lovemaking seems particularly strong among human beings, and in nearly all cultures.

The adaptive advantage of such behavior should be apparent. The deeply personal involvement so characteristic of human lovemaking—whether ventral-ventral or not—precludes a watchful eye for possible predators or competitors. Hence, early prehumans making love in an exposed savannah during broad daylight would be especially vulnerable, and less likely to rear successfully the eventual products of their intimacy. Secrecy would thus enhance safety, and success. Furthermore, mating is often physically demanding in our species, and both partners are generally less capable of vigorous physical exertion immediately following orgasm. This further jeopardizes the safety of public copulators, although of course, the same may hold true for other animals. But the intense orgasmic involvement of both sexes in humans exceeds that of all other animals and one might therefore expect the physical vulnerability following sexual intercourse to be greatest in human beings as well.

A final advantage to private sex derives from our general capacity for nonseasonal sex and the ease with which we can be stimulated. Thus, as Freud pointed out, human societies surround themselves with sex images and symbols reflecting both subliminal urges and conscious efforts at titillation. Highly sexed as we are, the sight (or sounds, or in fact, any reminder) of sexual activity is especially exciting to Homo sapiens and the public copulator runs the risk of stimulating immediate competition from among the onlookers.

It is not surprising that private sex is generally characteristic of modern human beings. Along with this clearly adaptive characteristic there have developed countless supportive cultural practices. One might expect this situation to represent another example of adaptive cultural characteristics, but then again, our culture and our biology do not perfectly complement each other. Within Western culture it is common for parents to overcompensate in defense of their sexual privacy, thereby imparting confusion to their children, and a feeling that sex is "dirty." Most Christian religions contain various sexual prohibitions, which may be seen as attempted defense of sexual privacy. But even among parents not strongly influenced by organized religion, a deep shyness often pervades and even prevents discussion of sex between parent and child. The result can be an adolescent and eventually an adult with a complex of guilt, confusion, and possible sexual inadequacy.

Another avenue to neurosis is provided by culturally mediated opportunities for massive overcompensation, resulting in exhibitionism, either through drugs or financial reward. In the latter case, a recent cartoon is especially significant: a naked man and woman are in bed being filmed for a movie scene, surrounded by lights, cameras, and countless people. The man, betraying an easily understood and clearly adaptive response, is complaining, "Somehow I just don't feel like it!"

Another basically adaptive characteristic of human sexual behavior that is ancient, widespread, and probably influenced by our genetic makeup (our "nature" as well as our "nurture"), is the so-called incest taboo. The horror of Oedipus may well be universal among human society. To understand the evolutionary advantage of prohibitions against matings of close relatives, we must first examine some basic facts of genetics.

Most genes occur in pairs, one set provided by the father and one by the mother. Both parents may contribute identical genes for a given trait or each may supply a different variant, in which case the offspring would be, in effect, a genetic mosaic for that particular characteristic. This does not necessarily result in an actual mosaic appearance for the characteristic itself; rather, one gene may "dominate" the other and be largely responsible for the final appearance of the trait, or the result may be somehow intermediate between either pure appearance, or even entirely different from each. Individuals carrying two differing genes for a particular trait tend to be healthier, more vigorous, and thus more fit than those possessing two identical genes. There are many possible reasons for this. One of the simplest is that disadvantageous genes arise from mutations and are generally recessive, not showing themselves in the trait because they are masked by their partner, which is then considered the dominant gene. In cases of genetic nonidentity, deleterious recessives can thus be masked by the corresponding dominant. But when both genes are identical, two recessives may occur together, and as a result the less advantageous recessive trait is expressed, and therefore the individual in question is less fit.

Considering the advantages of genetic variety, we would expect evolution to have devised numerous tricks for discouraging genetic identity among any pair of chromosomal traits within each individual, and one of the best ways to guard against such occurrences is to insist that mated pairs be related distantly, if at all, while prohibiting reproduction by close relatives. For example, imagine an individual containing identical genes for a particular trait; these genes may be represented by "GG." A relative would stand a good chance of being genetically the same (the chances would increase the closer the family relationship), thus producing "GG" offspring. A complete stranger, however, may have the genetic makeup "gg." Mating between them would thus produce the usually superior "Gg" offspring, one gene having been contributed by each parent.

Long before the genetic basis for this phenomenon was discovered, animal and plant breeders had known that continual mating of closely related individuals would eventually reduce the quality of the stock. "New blood," they knew, had to be brought in every once in awhile, otherwise the population would decline in health and vigor. Geneticists now term the deleterious results of extensive breeding among close relatives "inbreeding depression," something that is avoided among most animals by dispersal, the tendency of juveniles to leave home and seek their fortune elsewhere, away from their close genetic relatives. Among prehistoric human beings, however, dispersal was probably uncommon, since our species is so highly social that solitary, dispersing individuals would be unlikely to survive. So other mechanisms were necessary to prevent inbreeding. The common prohibition against mating of close relatives thus not only is a cultural practice with clearly adaptive biological significance, it may also be supported by a genetic tendency resulting from the direct pressures of natural selection upon our ancestors.

Israeli anthropologist Joseph Shepher has documented that children growing up in a kibbutz, who have been treated as though they were members of one large family, tend to avoid marrying each other, despite the fact that social pressures actually favor such unions. The young people themselves report that marrying a member of one's early play group would feel like "marrying a sister (brother)." Even though the individuals in this case are not biologically related, it seems likely that their underlying biological predispositions are inducing them to avoid "incest," using a social cue which is normally accurate but which, in the particular cultural environment of a kibbutz, is no longer reliable. Thus, by avoiding matings within one's play group, "biological" Homo sapiens has minimized the danger of incest and inbreeding, for many generations, whereas "cultural" Homo sapiens has outsmarted himself.

It may be objected that the incest taboo has not been absolutely universal among human beings. The well-known Egyptian pharaoh Tutankhamen and his wife, Queen Nefertiti, for example, were brother and sister. An extensive line of European royalty was plagued with hemophilia because of intermarriage among close relatives. Matings of first cousins have long been common in east European societies. Sociologist Pierre van den Berghe, among others, has pointed out that incestuous matings, especially between brother and sister, tend to occur in very special social situations: when enormous power, wealth, and/or spiritual values are to be concentrated within a very select circle. In such cases, cultural rules predominate over biological predisposition. (Also, at least the men often wind up having additional children via concubines, to whom they are not closely related; the royal women, by contrast, are more likely to be stuck in a situation that is, for them, biologically maladaptive.)

When biologists suggest that a trait is genetically influenced, they do not imply invariant, exact correspondence in all cases. We do not inherit blue eyes or tall stature as such, although both are under genetic influence. Rather, we inherit a potential, a range of possible expressions that will vary within certain limits, often rather wide limits at that, depending on the specific environment in which we live. Second-generation Japanese-Americans, with a gene pool that is indistinguishable from that of their parents, are nonetheless considerably taller than their parents because of better nutrition available to them in the United States. We may well have a biological predisposition for copulating with strangers, but certainly, we can meet strangers without copulating with them. And conversely, a biological predisposition for incest avoidance does not guarantee that incest will never occur.

To take another simple biological example, there is a genetically controlled characteristic in rabbits known as "Himalayan," in which the animals have patches of black on the tips of their forepaws, hind feet, and ears. The rest of the fur is pure white. The extremities of all animals are somewhat colder than their more central regions (witness our human problem of cold hands, feet, and ears) and rabbits possessing the Himalayan genes respond to cold temperatures by growing black fur. This can be experimentally tested by shaving off a patch of normal white fur from the side of a Himalayan rabbit. As expected, it grows back white again. Now, however, shave some fur and apply an ice pack during the regrowth period: the new fur comes in black. Clearly, black fur itself is not inherited in Himalayan rabbits, but rather the *ability* to produce black fur if the temperature falls below a certain threshold.

A more complex, but analogous example occurs in humans, with the "inheritance" of intelligence. There is clearly some genetic basis for differing IQ levels, but the possession of all the right genes for great brilliance will not make a genius out of a child raised in complete isolation, or one even moderately deprived of the necessary opportunities for mental exercise and growth. We inherit a range of possibilities, but the specific realization of these potentials depends on the particular environmental factors to which the individual is exposed. Similarly, a genetically mediated inclination for ventral, private copulation with nonrelatives and with a diversity of partners—perhaps more so for males than for females—could easily be swamped by cultural factors including religion, conscience, technology, and habit, with little regard to its ultimate biological utility.

When incest occurs in situations not involving royalty, it is most frequent between father and daughter, rare between siblings, and almost unheard of between mother and son. Moreover, by far the most frequent example of incest involves a stepfather with his stepdaughter, cases that are often closer to child abuse and that do not meet the biological definition of "incest" at all.

As we shall see in more detail in the following chapter, males and females are different, not only in their fundamental biology, but in certain behaviors as well. Not surprisingly, the behavioral differences are consistent with the more strictly biological. Males, in brief, are the inseminators; females, the inseminated. Males—as expected—are therefore the sexual aggressors and among many animal species, they are sometimes rapists as well. Through studies of ducks, geese, fish, and even insects, biologists have been compiling impressive catalogs of rape: the common factor seems to be that under certain conditions, notably when they are relatively unsuccessful otherwise, males attempt to maximize their fitness by forcing copulations with females.

A recent study by Randy and Nancy Thornhill has suggested that fitness considerations may be relevant to rape among human beings as well. They found that rape victims tend to be primarily women of childbearing age, whereas murder victims, by contrast, tend to be more evenly distributed across the population at large. If rape was simply a crime of violence against women, it would presumably be directed equally against all women. Moreover, the Thornhills also found, as predicted from evolutionary theory, that young men are overrepresented among rapists, and that these men tend to be relatively low in socioeconomic status, and thus, likely to be unattractive as mates if they followed socially sanctioned sexual strategies instead of rape.

Men are not only the most frequent transgressors in incest and rape, they are also far more likely than women to respond violently to adultery by their mate. In fact, the most frequent cause of violence among spouses, from wife-beating to murder, from Westchester to Mozambique, is sexual jealousy . . . especially male sexual jealousy. Among many societies, adultery is considered a crime only when it is performed by a wife, not by a husband. And moreover, it is often specifically designated a crime against the "offended husband," who may then, with the approval of society, wreak vengeance against his wife and/or her lover. There is little good evidence allowing us to compare female sexual jealousy with its male counterpart, but it is almost certain that among the great majority of human societies, women are much less enraged by their husbands' philandering than men are by the same behavior on the part of their wives. Much of this difference may be attributable to cultural traditions, which prescribe greater female tolerance and a sexual double standard. However, these cultural patterns, important as they are, also seem likely to emanate from biological patterns that are no less important: women get pregnant, not men. Accordingly, sexual dalliance by a wife is much more likely to compromise the fitness of her husband than vice versa.

With recent cultural advances in the technology of birth control as well as abortion, the strictly biological consequences of extramarital infidelity have been substantially diminished. As with the biological significance of the pair bond, the widespread cultural hyperextension of the double standard as well as male sexual jealousy therefore appears to be more inappropriate than ever. But not surprisingly, our "all too human" emotional responses—often intolerant, jealous, and sometimes even violent—still lag behind, caught in an evolutionary time-warp.

As intelligent, moralizing creatures, we can look at the seamier side of our own behavior, and establish cultural norms and codes that seek to outlaw and prevent actions that we judge to be intolerable. However, we do a disservice to the cause of humane understanding and ultimately, social betterment if we willfully ignore or deny the likely biological component of precisely those behaviors which we find so offensive. The likelihood that an act is in some sense "biological" does not make it good, just as being "cultural" does not make it bad. Few things are more "organic" than typhoid; when medical science seeks to understand the disease, this is not to support or encourage the typhoid bacillus. Similarly, when evolution offers some insights into rape and sexual violence, this is not to condone the behavior, but rather to help us understand something that outrages our cultural norms.

In the words of Saki (Hector Hugh Munro), "It takes all sorts to make sex," some sorts that we appreciate and enjoy, others that we find uncomfortable, embarrassing, and disagreeable, and yet others that are simply despicable. Underlying them, for better or worse, are all sorts of biology and all sorts of culture, variably dependent, sometimes opposed, and yet always intertwined.

CHAPTER FIVE

FEMINISM

Gorillas to Goldman to Gilligan

Marriage is primarily an economic arrangement, an insurance pact. It differs from the ordinary life insurance agreement only in that it is more binding, more exacting . . . If, however, woman's premium is a husband, she pays for it with her name, her privacy, her self-respect, her very life, "until death doth part." Moreover, the marriage insurance condemns her to life-long dependency, to parasitism, to complete uselessness, individual as well as social.

—Emma Goldman, "Marriage and Love" (1910)

Feminism is not unique to the 1980s or even the 1970s; its precursors can be identified at almost any stage of human history. What is unique, however, is its intensity, breadth of support, and depth of commitment in modern Western culture. It is clearly a cultural phenomenon, prominent in some societies, absent in others. It is almost certainly nongenetic in origin and is most commonly associated with a whole constellation of social and political values, in the Western world generally leftist and anti-establishment, since the Western "establishment" tends to be male dominated, politically conservative, and rather sexist. The correlation between political-social conservatism and sexism is so strong that it is often taken for granted. Here, as elsewhere, those things that we take for granted are most likely to yield insights into our innermost inclinations. In addition, it is not usually recognized that at the center of the feminist struggle—between woman and man, between woman and sexist society, and within many women themselves—is the persistent conflict between cultural and biological evolution.

Men and women are different, and not just in anatomy. Although societies may differ in what is recommended or even acceptable for either sex, and despite the fact that "men's work" in one society may be "women's work" in another, it remains true that men's behavior tends to differ consistently from women's behavior, in all human groups. In addition, a distinct pattern can be identified: women, for example, are universally more likely to be engaged in child care and men are universally more likely to go hunting and to kill each other. Whereas the evidence is accumulating that these differences are based at least to some degree on deep-seated biological distinctions between the sexes, we seem to have created cultural differences that go farther yet. Insofar as these differences are not commensurate with each other, we have a great source of irritation and injustice.

For some time, feminists resisted any efforts to identify differences between the sexes, and rightly so perhaps. Patronizing attitudes toward women as the "weaker sex," with implications of greater emotionality, lesser rationality, and generally less competence, have bolstered a wide array of institutions that have oppressed women for generations, and which to a large extent still do.* Locked in a vigorous struggle for equality, it seemed appropriate to resist as "sexist" any implication that women are not in every way "equal" to men. Early in any movement, it is convenient—often necessary—to simplify issues in the interest of clarity

*At a speech following the 1984 elections, Geraldine Ferraro pointed out to the great amusement of her (mostly female) audience that during their televised debate, Vice President Bush had shown dramatic mood swings, serious one moment then almost giddy the next. Ms. Ferraro wondered aloud whether this was hormonally based, and whether the nation was well advised to leave its direction to men, with their apparent biological instabilities.

and to prevent divisive disputation and destructive self-doubts. Fortunately, it seems that feminist thought has now matured, just as social and political thought generally tends to mature when its initial revolutionary goals begin to be achieved. It can therefore be hoped that feminists are now prepared to deal with, and to profit from, the insights of evolutionary biology.

Women have long been oppressed, in a variety of ways by most human cultures. Despite the conservative trend in the United States during the 1980s, it seems unlikely that women will much longer be denied equal opportunities, status, and respect, in both the workplace and the home, and equal pay for equal work. Like environmentalism before it, and perhaps the nuclear freeze movement in the near future, feminism has finally been transforming itself from a seemingly radical departure to an accepted way of thinking. In the process, it is at last becoming intellectually respectable to confront certain truths that were in fact always true, but were also inconvenient. Feminism as a legitimate claimant to social change will in the long run be strengthened by attention to biology and to the conflict between biological and cultural evolution, no less than by attention to economics and the conflict between private needs and male-dominated state policy.

Everyone has known for a very long time that men and women are structurally different, that women make eggs and men make sperm, that men have penises and women vaginas, that women have babies and lactate while men do not. These facts are not sexist; they are just facts. On the other hand, it may indeed be sexist to interpret some of these differences from a biased male perspective and suggest, as Freud did, that little girls perceive themselves as missing something that little boys have, suffering therefore from penis envy. One could as well suggest that when little boys grow up and get themselves covered with blood in fights or wars, as they often do in their sexist way, they are suffering from menstrual envy.

Aside from the obvious sexual differences, there is a clear-cut distinction between men and women when it comes to physical size and strength. Although there are exceptions, men tend to be larger and stronger, and these differences are due to biological rather than cultural evolution. This may seem surprising. Given the fundamental distinction in their biological roles, one might expect the females to be generally larger. After all, men supply the sperm; women, the egg. The man's contribution is transient and relatively trivial in mass and energy; the woman's is long lasting and immensely costly to her. Since it is the woman who must literally carry the biological weight of reproduction, it would seem logical for her to be sturdier than the man. In a sense, she is. Thus, although muscular strength belongs especially to the male, females are biologically stronger as shown by their almost universally longer life span.

On the other hand, male superiority in muscular strength is also a product of biological evolution. The reason? Defense and competition. If we look at a social monkey now inhabiting the same African savannah that was probably the birthplace of modern Homo sapiens, we can see some of the selective pressures at work that almost certainly acted in shaping us. The common baboons of east Africa live in social units consisting of many adult males, females, juveniles, and young. The males are very powerful and often twice the size of the females. They also possess enormous canine teeth and an aggressive temperament, both of which become apparent when one male challenges another for social dominance and mating rights, or when a predator such as a leopard or cheetah appears on the scene. In the former case, the adult males threaten and if necessary fight; in the latter, they often do the same thing.

A similar correlation between size and aggressiveness among males holds true for the other terrestrial primates, and in fact, for most mammals. Even among the "lordly" lions, in which, incidentally, the males virtually never exert themselves to make a kill, they still function in defense of the pride. In this case they protect the kill especially from appropriation by hyenas, which have been known to chase an unattended lioness from her hard-earned meal. In other respects, the male lion is a virtual parasite of the female, since she does all the hunting. But when it comes to defense of the group, especially defending his own status from other males, the male lion is no slouch.

Male lions compete vigorously with each other for "ownership" of a pride. To the victor goes the opportunity of copulating with several females, and thus, reproducing by them. To the losers, a resentful career as bachelors, periodically trying to achieve social, sexual, and thus, evolutionary success.

Defending the group is a dangerous occupation and it might seem logical that evolution should have chosen the relatively expendable male for the task, while preserving the more important female. However, so far as we know, evolution does not act to benefit the species or even the group as such; rather, whenever some individuals leave more offspring than others, or when some genes leave more copies of themselves than do others, natural selection is taking place. Any species benefit is therefore likely to be purely incidental to the evolutionary jousting of individuals and their genes. Analogously, a nation's Gross National Product is the incidental result of individuals, corporations, and governmental entities seeking to maximize their own functioning; the GNP will generally go up when each subnational component does as well as possible, but not because enhancement of the GNP is the goal or the level at which competition is taking place. Similarly, species benefit is the result of selection acting to maximize the performance of each lower level component, not enhancement of the species per se.

Male lions defend their group against others, both hyenas and other lions, because there is something in it for them: by doing so, they are more likely to be evolutionarily successful. Presumably, the same fundamental pressures acted for thousands of generations on primitive Homo sapiens as well. Given the biology of sperm-making versus egg-making, it seems likely that such defense was primarily defense of the individual male's reproductive success; that is, it was probably self-serving, and in no way particularly laudable, gentlemanly, or gallant.

Unless females actively competed for males, there would be no equivalent selection pressure operating on them. If, on the other hand, there was a shortage of females, competition would ensue with natural selection favoring those females whose characteristics help them secure a mate and breed successfully. Among the common North American shorebirds known as spotted sandpipers, the usual pattern of sexual politics is reversed; females are notably malelike. Larger and more aggressive than their mates, female spotted sandpipers arrive on their breeding territories while the males are still loitering down south. These pushy, competitive females fight it out with each other, after which the males arrive and demurely settle down on one territory or another, typically several males within each female's "harem." Each of these males is subordinate to his territory-owning female, and each one incubates a clutch of eggs laid by the dominant female, who busies herself defending her boundaries and attending to the world of affairs while he cares for the offspring, unaided.

A similar pattern is found among those peculiar creatures, the sea horses, in which the females transfer their eggs to the males, who incubate them in special brood pouches. Significantly, female sea horses—like female spotted sandpipers—tend to be larger, brightly colored, and aggressive, while the males are small, drab, retiring, and sexually coy. These are all exceptions, although interesting exceptions to be sure, since they help prove the rule that the sex investing more (generally females) tends to be less aggressive and to be the subject of sexual competition on the part of the sex investing less (generally the males).

Among some animals, success in sexual competition may well involve physical strength. Among others, such as human beings, traits of little or no immediate survival value may nonetheless be of reproductive advantage if they are attractive to members of the opposite sex. There is much current debate among evolutionary biologists as to whether such apparently nonadaptive traits really are preferred, and if so, whether they are really nonadaptive. For example, it is at least possible that the large antlers of deer are a liability rather than an asset, since they require a substantial metabolic investment, and may make their bearers more conspicuous to predators. On the other hand, if an impressive rack of antlers indicates that its carrier is capable of surviving despite such a handicap, then females may preferentially mate with such an individual . . . thereby rendering the "handicap" an advantage! In any case, there are often strong selective pressures favoring large, impressive, and aggressive males who can defeat or dominate other males, while defending themselves and their dependents against attack.

Something similar to this phenomenon was first noted by Charles Darwin, who named it "sexual selection," mistakenly believing that because it often depended on the choice of attractive mating partners by the females—who were somehow possessed of an intuitive aesthetic sense—it was quite different from *natural* selection. We now recognize that selection is indeed occurring in these cases, but the selecting agent is the entire environment of the species, not just the proclivities of its females. Males do indeed strive to impress females, but perhaps even more, they attempt to outcompete other animals and particularly, other males of the same species. When the dominant silver-backed adult male gorilla repels the advances of another silver-back from another troop, he succeeds not only in male-male competition, but also in assuring himself the sexual attentions of his females.

Again, it might seem that evolution would still have produced larger and more powerful females by this same mechanism: females adept at defending the group from predators and capable of defeating possible competitors from other groups would naturally leave a greater number of successful offspring. This would cause selection for greater size and strength ("male characteristics") among females as well. And to some extent this has happened. But the fundamental biological distinction between male and female would foil the extensive development of female Amazonism. Thus, although a highly successful male could make a major contribution to the next generation's genetic makeup, the contribution of even a very successful female is limited by her biology. Reproduction in mammals requires a lengthy pregnancy, resulting in only a few offspring at a time; in primates it usually requires an even longer postnatal period of maternal care and attention. So, even though individual females, more than individual males, determine the reproductive output and ultimately the evolutionary success or failure of their species, even a hypothetical superfemale can make at best only a small genetic contribution to the next generation. By contrast, a successful male can fertilize many females, and therefore have a much greater evolutionary impact. The upshot of all this is a further tendency toward physical as well as behavioral differentiation between men and women among our ancestors.

This does not imply that males on balance have an advantage over females. In fact, the occasional male that is wildly successful is exactly balanced by many males who are not. Females, by contrast, are less likely to strike it rich via natural selection, but they are also less likely to strike out altogether. Being a male, then, is inherently risky, since among many species males are either successes or failures; being a female, by contrast, is more conservative, since nearly all females will breed, and the disparity between success and failure, biologically measured, is usually not very great.

But nothing in biology is gained for nothing. Most advantages are obtained at the expense of some disadvantages: A larger, stronger body requires more nourishment, and in times of food shortage, it is undoubtedly advantageous to have less bulk to support. This would provide a compensating advantage to smallness. Moreover, brightly colored males are necessarily more conspicuous, and thus, more susceptible to predation. And finally, whereas dominant males may be successful in evolutionary terms, they typically have a shorter life span than the evolutionary failures, which spend less time fighting and copulating, and more time eating and keeping out of harm's way. A mountain sheep ram, for example, or a bull elk, may well be exhausted and emaciated at the end of the rut, especially if he has been successful in maintaining his harem. Not surprisingly, females live longer than males among most animals, just as women live longer than men.

As a consequence of natural selection acting somewhat differently on males and females, the two sexes are therefore somewhat different, both physically and behaviorally. These biological differences also influence not only the proclivities of each sex, taken separately, but also the ways in which males and females, men and women, interact with each other. Males clearly dominate females among the monkeys and apes. Gorilla bands are led by a dominant silver-backed male; a clear progression of rank among dominant males can be identified among rhesus monkey bands and an oligarchy of several males often rules baboon troops. Frequently, the females have a distinct social hierarchy all their own, but any randomly chosen female is generally subordinate to any randomly chosen male. In addition, whereas male hierarchies are usually stable, changing only rarely when high-ranking members die or get very old, the female social ranks are constantly shifting, as a result of changes in the estrous cycle, presence or absence of dependent young, and possible consort association with a high-ranking male. Such biologically generated instability makes female dominance over males highly unlikely. In no cases, therefore, are primate groups consistently dominated by females. This is almost certainly a direct result of the biological distinctions between the sexes, based on hormonal and anatomical differences, which in turn are attributable to selection for male success in defense and competition, and for female success in cooperation and nurturance.

The characteristics of physique and temperament that make for aggressive success and were likely the object of intense selection among our ancestors seem to have generally led to social dominance of females by males. This correlation is not 100 percent, but common experience reveals how often domination goes along with physical size, strength, and aggressiveness. It is certainly true in most animals, and human beings do not appear to be an exception.

Male dominance in human beings might thus seem to be the biologically "correct" state of affairs. If we were purely biological beings, the issue would be closed, and in fact, would never have been raised! But we are unique in serving two, sometimes rather disparate masters, biology *and* culture, and this changes things altogether. Let us grant, if only for the sake of argument, that the general dominance of men over women derives from physical and behavioral differences due largely to selection for male aggressiveness, competitiveness, and defense of the social unit. There was probably great need for such characteristics as recently (in evolutionary terms) as 30,000 years ago. Since that time, however, cultural evolution has completely remodeled the contingencies of human survival and social life, with biology, as usual, lagging far behind. Men today only rarely function in direct defense of their genetic legacy; given the presence of nuclear weapons, in fact, those who do participate in such "defense" and who urge it upon the rest of us may ironically be endangering not only themselves but also those they claim to defend. They may in fact be the greatest threats to long-term evolutionary survival that our species has ever confronted.

The human tendency for a "nuclear family" with a strong male-female pair bond further reduces sexual selection for enhanced male characteristics. Thus, with roughly equal numbers of males and females in the population, and with (culturally imposed) monogamy, virtually every male gets to reproduce, instead of a few extraordinarily endowed specimens doing the lion's share of the breeding. There seems little doubt that given human biology and culture, selection for male-female differences will be decreasing. But there is equally little doubt that given the rate of biological evolution, such biological equalization must probably be measured in thousands of years. Where does that leave us?

We are stuck, saddled with an outdated biological system, rendered anachronistic by our rapidly evolving culture, and unacceptable by our expanding social consciousness. We possess physical characteristics and behaviors such as male dominance and aggressiveness that are offensive, inappropriate, and often downright dangerous. Although human culture, ironically, is itself responsible for this discrepancy, much of the difficulty in readjusting society stems from the cultural—not biological—supports to male dominance that have been erected by men. Many traditional behaviors, including such diverse things as retaining surname at marriage, identification of most political and economic activity as a male province, expected man-woman roles in home maintenance and wage earning, smoking cigars, or access to seats on a crowded bus, are purely social conventions. Admittedly, they may have been largely developed in response to the biological tendency for male protectiveness, competitiveness, and resulting patterns of dominance. But, such dominance is now irrelevant. Combined with widespread awareness of this fact, as well as the (culturally mediated) flexibility of human behavior, we can look forward to their elimination. But at the same time, cultural hyperextension of female suppression will not be dismantled quickly or easily. It may also require a hyperextended awareness of the need for such remedies.

There is a delicate titration between biological predispositions and social prescriptions. In such cases as male-female differences, it seems especially likely that society, seizing on a degree of evolutionary reality, has hyperextended it, making awkward, unjust, and offensive cultural mountains out of what may, in essence, be biological molehills. Men and women are indeed different, but in most cases they are less different than human social traditions have demanded them to be. Give cultural evolution a hand in such cases, and it takes the whole arm, ultimately, perhaps, to the detriment of all concerned. In other situations, the opposite may occur: evolutionarily given realities may be diminished rather than augmented by the action of cultural evolution. When this involves socialization to reduce otherwise potentially abrasive consequences (for example, of male aggressiveness), this cultural sandpaper may be quite beneficial, so long as it is applied gently and with sensitivity, as well as persistently. In other cases, cultural practices may run directly counter to evolutionary inclinations, and the result, while presumably in accord with conscious intent, may nonetheless cause great distress.

When it comes to male-female role differentiation, some biology remains clear. Among all mammals, females get pregnant, not males, and moreover they are equipped with milk-producing glands by which the infants are nourished after birth. Nearly all female mammals also possess genetically influenced behavior patterns that provide for care of the young. The new mother must generally remove the fetal membranes, usually eating them for the added nourishment and/or hormones they provide. Among many species, the newborn must be licked profusely in order for normal urination and defecation to occur. Adults often brood their young, warming them if they are cold, shading them if they are hot, removing fecal pellets to prevent fouling of the nest, and providing a steady supply of food. Such provisioning may or may not involve assistance by the father, but it always involves nursing by the mother. Female rats and mice will automatically retrieve their young if they have been scattered about, collecting them in a pile where they will be kept warm and safe. In most cases, this complex of maternal behaviors is activated by hormones produced by the mother herself, particularly the hormones oxytocin and prolactin, associated with lactation.

The situation in humans is undoubtedly more complex than among any animal. A good generalization in the study of animal behavior, in fact, is that among progressively smarter animals, the role of genetically influenced and hormonally activated behavior declines as the cerebral component increases. We have already seen, for example, that sexual behavior in human beings has been liberated from the imperious chemical domination by hormonal cycles that is found in all other mammals. Unquestionably, maternal behavior in Homo sapiens has also been greatly emancipated from the control often exerted by hormones and genetics. There is, in fact, no good evidence that mothers who are bottle feeding their infants have weaker maternal drives than do breast feeding mothers, in whom the maternal hormones are more prominent.

However, breast feeding has an undeniable side effect on the nursing mother's physiology: it tends to inhibit the next ovulatory cycle. This so-called lactational amenorrhea is entirely biological and is adaptive in reducing the likelihood that a mother will become pregnant again while still nursing an infant. In fact, among nontechnological societies, there is a correlation between the length of normal lactation, the length of postpartum sex taboos (socially mandated restrictions on sexual intercourse following the birth of a child), and the availability of protein in the maternal diet: lower protein levels tend to be associated with longer nursing and other culturally elaborated practices that reduce the probability of pregnancies following, maladaptively, too close one after the other. When corporations selling artificial infant "formula"—such as Nestlé in the recent past—successfully market their product in Third World nations, they disrupt this delicate synchrony between biology and culture, thereby removing an adaptive inhibition against excessive, unwanted pregnancies. In addition, whereas mothers' milk is generally free of pathogens, artificial infant formula made with local, contaminated water contributes significantly to chronic diarrhea and dysentery, major causes of death among newborns.

There is no unimpeachable evidence that childcare and nurturance must necessarily be women's work rather than men's. There is, in fact, overwhelming evidence that men are capable of substantial, and effective "mothering." The expression of any particular nurturant behavior is obviously very susceptible to cultural influences, especially social expectations and economic situation, as well as personal preferences, often born of one's own idiosyncratic experiences. However, it seems unlikely that a behavioral system so important to evolution as care of the young would not have some genetic component, even in that most liberated of species, the human animal. The system may be flexible and easily influenced by culture, but something is almost certainly there nonetheless. The well-known sensitivity of menstrual cycles to stress as well as the subtle and problematic but nonetheless likely effect of these cycles on behavior suggest that Homo sapiens is not immune to the effects of biology, in this case, our reproductive hormones.

Women's fear of identifying behavioral differences between themselves and men, combined with men's rather shortsighted and sometimes even churlish insistence that "men" somehow represent the entire human species, has resulted in some limited views of normal human nature. In her justly influential book *In a Different Voice: Psychological Theory and Women's Development,* Harvard professor Carol Gilligan has effectively outlined some of these misunderstandings. For example, girls have traditionally been thought to lag behind boys in the most widely acknowledged measures of moral development, devised by (male) psychologist Lawrence Kohlberg. This is because boys tend to evaluate situations by reference to abstract laws and ethical principles, whereas girls tend to focus on social connections and interpersonal relationships. There is, moreover, no arbitrary standard by which we can judge the former relative to the latter, although an appreciation of evolutionary biology can help us understand why the former is primarily a male pattern and the latter, female.

Gilligan emphasizes that female moral preferences tend to place "an ethic of responsibility as the center of women's moral concern, anchoring the self in a world of relationships and giving rise to activities of care," whereas by contrast, the highest stage of moral development, according to accepted psychological dogma, favors recognition of universal rights rather than personal responsibility. "The morality of rights," as Gilligan points out, "differs from the morality of responsibility in its consideration of the individual rather than the relationship as primary." It is probably no coincidence that a morality of rights rather than responsibility also happens to be the typical moral preference of males, and that the designers and interpreters of these systems also happen to be predominantly men.

Developmental psychologists have noted that when children

are playing, boys' games tend to last longer than girls', because boys quickly learn to adjudicate disputes by reference to abstract rules and principles. Gilligan points out that

> By participating in controlled and socially approved competitive situations, they learn to deal with competition in a relatively forthright manner—to play with their enemies and to compete with their friends—all in accordance with the rules of the game. In contrast, girls' play tends to occur in smaller, more intimate groups, often the best-friend dyad, and in private places. This play replicates the social pattern of primary human relationships in that its organization is more cooperative.

By contrast to boys, girls are more likely to stop the game when a disagreement arises, because they value the relationship among the participants more than abstract, blind justice, or the game itself. Although Gilligan refrains from suggesting the origin of the difference she so eloquently describes, its compatibility with biological evolution suggests that natural selection has been involved.

The biological evolutionary task of both males and females is to succeed in projecting copies of their genes into the future, to maximize their fitness. But as we have seen, success is likely to be achieved somewhat differently among the two sexes. Male success is typically achieved by effective competition; female success, by relationship, especially with their own offspring and other relatives. Thus, for boys and men, morality is at its most ideal and alluring when it is a morality of justice, of theoretical principles that place restraints upon aggressive, competitive, self-serving tendencies; for girls and women, on the other hand, morality is suffused with images of relationship, of caring, and of taking care of others. Male morality, as Gilligan describes it, is an ethic of *inhibiting* one's nasty self; female morality, in contrast, emphasizes *releasing* the caring self.

The classic developmental task for young boys, recognized by psychiatrists and psychoanalysts, is differentiation and individuation. It is a task that is appropriate to their biological task as well. Boys must separate themselves, physically and emotionally, from their primary caretaker (who is typically the mother), and become something different: a man, and a father. Becoming "your own man" means becoming different from your mommy. Young girls, however, are behaving more in concert with their ultimate biological needs if they model after their mother and achieve comparable relationships. In a world oriented toward achieving separateness and individuality, attachments appear as hindrances, as impediments to maturity. Erik Erikson has suggested similarly that for boys, identity must precede intimacy, whereas for girls, identity is found through relationships with others.

As Gilligan points out: "Since masculinity is defined through separation while femininity is defined through attachment, male gender identity is threatened by intimacy while female gender identity is threatened by separation. Thus males tend to have difficulty with relationships, while females tend to have problems with individuation." True to their biologically appropriate roles, men orient themselves toward career and success in the competitive world, entering readily into a series of hierarchically arranged systems, while women, true to theirs, orient themselves within networks of relationships. So when Freud identified the highest goals of mental health as "the ability to love and to work," he was not, in fact, describing a dual accomplishment to which both sexes equally aspire. Whereas men seek to be alone at the top, fearing in turn that others might get too close, women seek to be embedded in a network of human relationship and fear being isolated. As Gilligan puts it:

> The images of hierarchy and web . . . convey different ways of structuring relationships and are associated with different views of morality and self . . . As the top of the hierarchy becomes the edge of the web and as the center of a network of connection becomes the middle of a hierarchical

> progression, each image marks as dangerous the place which the other defines as safe.

Men fear failure; many women fear success. More than a century ago, suffragette Elizabeth Cady Stanton was so frustrated with women's proclivity for nurturance and sacrifice rather than assertion and worldly accomplishment that she urged a radical realignment in women's values. "Put it down in capital letters," she told a reporter, "SELF-DEVELOPMENT IS A HIGHER DUTY THAN SELF-SACRIFICE." Gilligan's work helps us understand that for women, self-development requires more effort than does self-sacrifice, and a look at our evolutionary history helps us understand why.

Following Stanton's injunction, women in growing numbers have finally begun to seek self-development, and to chafe at the slow pace at which society has permitted them to do their duty. Angered by the unresponsiveness of their own societies, many women found themselves in conflict with resistant and rigidified male-dominated and male-oriented social and governmental structures. Women began to seek and obtain freedom from sexist roles and from their inferior social status, roles and status that seem to have been originally suggested by our biology but were greatly magnified by our culture. At the same time, they were embroiled in another conflict, this one within themselves, between human biology and culture. Even with the heartlessness and social insensitivity that has characterized American federal politics in the 1980s, the fact is that women are more liberated from the tyranny of their biology than at any time in the long history of our species. Culture can now provide virtually all the minimal requirements for successful child-rearing—baby sitters, clinics, day care centers, schools—making the woman's contribution less and less necessary.

We have therefore gone beyond the simple biological concerns of the male as hunter, protector, selfish competitor, and (sometimes) provider, and the female as gardener, forager, sexual object, nursemaid, and educator. Women therefore find themselves caught in a double-bind between opportunities and desires for independence on the one hand, and the old biologically inspired yearnings for the reproductive role that evolution has already mapped out—and upon which society has built, and men capitalized—on the other. Small wonder women are ambivalent about their roles and their lives.

Men, by contrast, are generally having an easier time, since the old qualities of aggressiveness and daring, once so useful on the savannah, can be transferred more readily to the "outside" world of work, business, and professional competition. The crunch of culture upon biology therefore makes itself felt in every woman who debates the merits of career versus motherhood and in every man who finds himself called upon to relinquish some of the outmoded perquisites of dominance.

There can be no doubt that marriage, as it has been practiced by the great majority of human beings for most of our biological and cultural history, has been an institution that oppressed women. There can also be no doubt that at least in the West, some changes are under way. It would be ironic, though, if in the process of realizing their true potential as human beings at last, women find that rather than divesting themselves from the yoke of domestic slavery, they are simply permitted to assume new responsibilities—in the marketplace of work and competition—while still being burdened by the same old biological baggage as before. In her book *The Hearts of Men,* Barbara Ehrenreich, no less than Emma Goldman in the epigraph to this chapter, recognizes the shortcomings of marriage based on male dominance, female dependency, and economic bondage. Yet she also points out the tendency of men to take advantage of feminist striving for liberation by liberating themselves from husbandly and paternal responsibility and relationship, following perhaps their own untrammeled biology. This has left women stuck with *their* biology and simultaneously deprived of the culturally constructed "safety net," which, despite its shortcomings, has been one of traditional marriage's saving graces. Maybe its only saving grace.

As men have become increasingly aware of the possibilities of their own liberation, and of the debilitating stresses—emotional, cardiac, ulcer-inducing—of being success objects, women find themselves under attack from a new direction, caught between the Scylla of marital dependency and the Charybdis of second-class citizenship in the workplace (with childcare and family responsibilities often undiminished). Like Gilligan, Ehrenreich shies away from underlying causes. But the flight of American men from marital and paternal commitment that she so deplores and that growing numbers of men find so attractive is also a flight from a culturally imposed system to a seeming Shangri-La of self-realization, made all the more tantalizing by its congruence with male biology. It may be true, as Dr. Helen Caldicott likes to point out, that the woman most in need of liberation is the woman within every man, forced to play out a difficult, stressful, and often downright unpleasant macho role. But there is also a biological male within every man, and he, at least, may well resonate with the prospects of liberation from monogamy and the commitment to wife and children that it usually involves. After all, polygyny is biologically "natural" to our species, only inhibited by cultural proscriptions. (Whether is it "right" is another question.) And another biological consequence of maleness as opposed to femaleness is that whereas women are guaranteed that their children are biologically theirs, men are not. For the vast majority of mammals, male fitness is achieved by consorting with as many females as possible, while offering little or nothing in the way of paternal assistance.

GREAT blue herons are large, regal-looking birds that inhabit marshes, eat fish, and mate monogamously. Douglas Mock of the University of Oklahoma has found that within a minute after his mate leaves the nest area to begin foraging, the male great blue heron begins courting other females. These "extramarital" courtship efforts are not in any way disreputable, and certainly they are consistent with great blue heron biology: occasionally, a female finds a better partner and deserts her mate, and when and if that happens, the first male is better off if he has a replacement waiting in the wings (or, on the wing). Great blue herons—both male and female—apparently feel very little personal commitment to each other, but substantial commitment to their personal biological success. By contrast, extramarital endeavors among human beings may or may not be disreputable (depending on our cultural evaluation of such behavior), just as they may or may not be fitness enhancing.

Female stickleback fish deposit their eggs in a nest constructed by the male. The females then swim away, entirely free of domestic responsibilities, leaving their mates to guard the young. Once again, biological femaleness and maleness, operating in the context of each species' particular situation, dictate clear-cut roles for these animals; there is no such thing as a "liberated" stickleback fish, an oppressed one, or any that are notably ambivalent.

We, no less than the great blue heron or the stickleback fish, are bequeathed our maleness and femaleness by biology, and there is little that we can do about it. Manliness and womanliness, on the other hand, are judgments rendered by our cultures and by ourselves, and here, presumably, there is much that we can do. Whereas maleness among great blue herons or sticklebacks leads in a straightforward, uncomplicated way to male behavior, maleness among human beings need not necessarily lead to manliness, any more than femaleness always produces womanliness. Society, no less than biology, dictates manliness and womanliness, and such expectations may or may not be paralleled by our own inclinations. Sometimes the transition is less than smooth, because sometimes the perceptions, expectations, and restrictions of culture do not accord with those of biology. Out of this crucible of conflict arise some of our most frustrating impasses, and some of the most exciting opportunities to define ourselves as not only fully sexual but also fully liberated, committed to ourselves but also to each other, and as a result, fully human no less than other animals can be fully stickleback, or fully heron.

CHAPTER SIX

OF FAMILY AND FRIENDS

Altruistic Genes Playing Selfish Games

A hen is just an egg's way of making more eggs.
—Samuel Butler

In a sense, evolution is terribly selfish. Since selection favors characteristics that lead to differential reproduction, it favors behavior that helps each individual at the expense of all others. We can assume that in most cases at least, there is "room" in the environment for only a limited number of individuals of each species, or similarly, room for only a limited number of genes, competing for chromosomal space. An individual may thus promote his or her success not only by acquiring characteristics that are personally advantageous, but also by actively discouraging the success of competitors; insofar as life is a "zero-sum game," that means just about everyone else.

Seemingly opposed to this logic is the simple fact that many animals are not loners, fiercely competing with all others for a place in the evolutionary sun. Rather, they are social and often quite cooperative. Aristotle suggested that man is also a political animal, by which he did not mean that we are instinctively Democrats or Republicans, but rather, that we insist on associating with others, sometimes for cooperation, sometimes for competition, but nearly always by choice. Yet, there is other ancient wisdom on this score, notably *Homo homini lupus* (Man is a wolf to other men). We must also note, however, that wolves hunt in well-organized packs, within which each wolf has its place. Honeybees construct elaborate hives that may easily house thousands of workers, bison live in large herds, and baboons enjoy a complex social life based entirely on group membership. Animal sociableness, however, is actually not quite the anomaly that it may seem. Each individual within these species is better off as part of a group than it would be alone. There are no solitary baboons, or at least not for long. Although a single baboon is relatively defenseless, an organized troop can hold its own, making even a leopard change its mind and look for easier prey.

In addition to group defense, many animals gain added alertness to a predator's approach by enlisting the eyes, ears, and noses of other group members. Some animals, such as termites, actually produce a particular environment (within the nest) that is more conducive to survival than anything that could be made by a solitary termite. Social life also provides greater opportunity for learning and for the passing of traditions, as when members of a rat pack learn to avoid poisoned bait because the leader avoids it. In short, there are many advantages to cooperative behavior and thus, many reasons why animals live in apparent social harmony. In such cases, the social life is of selective advantage to each individual who lives in this manner, because it increases his or her personal chances of survival and ultimately, reproduction. Evolution will thus favor seemingly selfless behavior, but for selfish reasons.

This analysis becomes more revealing of the human situation when we consider behavior that helps another but does nothing to enhance the survival of the individuals doing the act. Perhaps the most obvious example would be parental behavior. Reproduction is so common in nature that we tend to take it for granted. What possible selfish advantage accrues to the dutiful progenitor?

In terms of personal survival, the answer, nearly always, is "none." Because sexual reproduction requires that two different individuals get together, intimately, it may be a time-consuming, difficult, and sometimes even dangerous activity. Thus, the female black widow spider is so named because of her occasional and unsavory habit of devouring her mate. Praying mantises sometimes do the same (once again, the female consuming the male) and in this case, there is even some evidence that the male copulates more vigorously after he has "lost his head" over a female—the cerebral ganglia tend to inhibit reflexes encoded in the lower nervous system, so a decorticate male, beheaded by his mate who may be more hungry for protein than for sex, will sometimes copulate successfully with his executioner.

Animals will often expend considerable effort, and run substantial risks, in order to reproduce: Pacific salmon batter themselves against rocks and swim upstream many miles, exhausting and often killing themselves even before they reach their spawning areas. Red-winged blackbirds stake out territories in the spring, and then fight fiercely to defend them—not for a place to eat, or to sleep, but to reproduce. In virtually every way, their lives would be simpler, safer, and probably longer if they simply looked after themselves and didn't bother trying to breed.

The antics of courting and copulating animals are notorious, and conspicuous. Roosters crow, elk bugle, elephants roar, and whales sing their eerie arias, sometimes for hours at a time. At such occasions, it seems that love-starved Romeos and Juliets are especially likely to be noticed by their food-starved predators. One of the most dramatic examples of the cost of reproducing comes from the common North American fireflies. These animals use their flashing patterns to announce their species and sex, and attract a mate; different species use different patterns, each one a distinctive Morse code. Thus, the male flashes his identity, the female responds with hers, and after a number of reciprocal reassurances, they fly away together into the moonlight. But in at least one species of firefly, the sexual code has been broken, and by a predator: in firefly courtship, three is definitely a crowd. This unwanted third party gives the flashing pattern used by the female of another species, whereupon a male of that species is attracted . . . and ends up being a meal rather than a mate. Entomologist James Lloyd of the University of Florida, who discovered this lethal intrigue, refers to "firefly femmes fatales."

Even once the various hurdles are surmounted and reproduction is well under way, the attentive parent often expends considerable energy raising its offspring: certain warblers have been observed to make 1000 trips per day, bringing insect food to their hungry nestlings. Carnivores such as wolves must do extra hunting to feed young pups. Nursing mothers of all species require extra food in order to support their dependent young. Furthermore, by consuming a portion of whatever food is available, the infants become potential competitors with their parents during times of food scarcity or drought and not only that, but when they grow up they are likely to compete even more seriously.

In addition to experiencing food and energy problems, the parent—just like the courter or copulator—may be more susceptible to predators than an "unattached" individual. Pregnant females, especially just before giving birth, are slower and less capable of outrunning their enemies. Many animals, including all primates, carry their newborn young, and the burden of heavy, dependent passengers could make a crucial difference if escape is called for. Given all these drawbacks to parenthood, how could it have been selected for? In a sense, the answer is obvious: if individuals declined to reproduce, then their species would last no longer than the life span of those individuals currently alive. But this is not in itself an explanation of why reproductive behavior occurs. Individuals do not perform with an eye on the ultimate good of the species (human beings, on occasion, may be exceptions here). If reproduction ceased because of immediate advantage obtained by the nonreproducers, the species would indeed soon go extinct. But the possibility of extinction cannot influence the evolutionary strategies of living things, since there is no way that organisms can reach ahead into the potential future, and then modify their current behavior accordingly.

But animals try hard to reproduce, despite the disadvantages it may entail. They do so because they are the descendants of others who did so, and moreover, who succeeded. Although bodies may be altruistic in a sense, looking out for other bodies—notably, those special bodies we call children—genes are "selfish" in that they look out for themselves. In another sense, however, they are "altruistic" too, in that they look out for other genes so long as those other genes are copies of themselves, just temporarily housed in another's body. Only those individuals with such altruistic propensities reproduced themselves in the past, and that's why every living thing that is here is here, and also why such things seek to reproduce.

Technically, individuals are not selected for, genes are; and by producing offspring as carriers of the parents' genetic material, living things enter the evolutionary arena. Since genes are being selected, not individuals, reproductive and parental behavior would be highly advantageous if it led to a maximum number of successful offspring. Behaviors that result in leaving more offspring would be selected for, and selection would cause a greater number of those genes (in this case, for effective parental behavior) to be represented in the future population. Animals will therefore expose themselves to all the rigors, dangers, and—at least in humans—irritations and inconveniences of having children in order to reproduce their genes. Not only will animals reproduce, but many will submit to considerable risk if necessary, in rearing and defending their offspring. The leaving of successful offspring is "worth" the chance of personal disaster.

This is not to suggest, incidentally, that living things seek to reproduce because of a conscious desire to bask in the warmth of natural selection's smile. Rather, selection has produced a host of short-term gratifications that lead animals to reproduce. Finding a mate or a nest site, copulating, caring for young, etc., all become, in effect, little gratifications in themselves. Animals go about satisfying one short-term need after the other, often being reinforced in the psychological sense, scratching the various itches that natural selection has established and that ultimately lead to selective advantage. It is not necessary for a reproducing animal to be aware of reproduction as its ultimate goal, any more than it is necessary for a tree to know that it will be more successful if it flowers in the spring instead of in the fall. Producing a maximum number of surviving genetic replicates is the ultimate goal, whether living things know it or not, and indeed, they can do a credible job of achieving their goal regardless of whether or not they are capable of acknowledging or even conceiving of its existence. Similarly, even a beginner can play a surprisingly good game of chess if he or she concentrates on the various possible subroutines—threatening two pieces simultaneously with a knight, containing the opposing queen, etc.—without necessarily orienting every move toward the ultimate goal of capturing the other side's king.

There is a limit to such a system. Although parents of many species will generally defend their young, a strategic retreat is usually preferred to certain death when the outcome is otherwise unavoidable for both parents and young. In such cases, individuals can ultimately leave more offspring by giving up on the present brood and escaping to raise another one. He or she who breeds and runs away may live to breed another day.

But, natural situations are rarely so cut and dried as this. Defense of a litter may or may not be successful, depending on such variable factors as the nature of the attacker, intensity of its assault, age and condition of the defenders, and degree of development of the young. A nesting pair of blackbirds will defend their young against marauding blue jays, but beat a hasty retreat before an approaching hawk. Each situation presents a different probability of success and hence, each must be assessed separately by the defending parent. This evaluation need not be conscious and rational, however. For example, marsupials differ from mammals in that their young are born in a very immature state, after which they develop in the mother's marsupium, or pouch.

One consequence of this arrangement is that the actual nourishment of marsupials is less efficient than in the case of most mammals, whose placenta permits a more effective exchange of nutrients. But there are some advantages for the marsupial as well: when chased by a predator, some species of wallabies and kangaroos have actually been known to abort a relatively large joey, thereby increasing the probability that the mother—now significantly lighter—will be able to run away. By contrast, placental mammals can hardly pause in mid-chase, undergo a convenient abortion, then keep running. Biologists speak of marsupial and placental mammals as each practicing a different reproductive "strategy," which simply means that they go about obtaining their evolutionary success in different, well-organized ways, whether they realize it or not.

The reproductive future of the adult also seems to be included in the evolutionary calculus. Thus, individuals who will not reproduce again or for whom each offspring represents a great investment of time and energy, will be more likely to risk all in defense of their young. Less persistence is expected from parents with a long reproductive life ahead of them or "easy breeders" with a negligible investment in any particular offspring.

Admittedly, it may seem cold hearted and insufferably clinical to interpret parental love in such "materialistic" evolutionary terms. But there can be no question of its validity among animals, and human beings, while very special, are very much animals as well.

Among the higher animals such as ourselves, hormones and genes are increasingly supplemented, modified, and even replaced by mental control of behavior. A female rat, deprived of the hormone prolactin, will not nurse and care for her babies. A female human being, similarly deprived, will not lactate, but can do a perfectly good job with a bottle; unlike the case with lower mammals, there is no evidence that maternal tendencies among Homo sapiens are evoked by maternal hormones. Our brain does the trick. Similarly with sexual receptivity, aggressive behavior, and so forth: among human beings, the highly developed brain makes decisions about behavior that among lower animals is controlled almost automatically by hormones and instincts. Not surprisingly, therefore, learning—and lots of it—is essential in our species. For the brain to be in control, alternative courses of action must be stored, this information must be readily retrievable, and there must be some ability to modify responses depending on what worked in similar situations in the past. If we relied on hormones and genes, we could simply trust our success to preconstructed chemical messengers, acting in concert with prewired neural circuits. But because our evolutionary strategy emphasizes a brainy flexibility, we are obliged to feed this brain, and not just with nutrients. And this provisioning of experience must begin early in life.

Prolonged dependence of the young is therefore the rule. The extraordinarily long period of infant helplessness among human beings requires continuous attention from the parents. During this time, children must not only be taught the basic requirements of successful living, as well as be defended from possible enemies, they must also be fed, cleaned, and cared for. The lower animals meet these requirements by relatively simple, automatic behaviors. Many birds respond instinctively to nestlings gaping for food. Students of animal behavior have found that many animals are automatically sensitive to certain—often exaggerated—signals in their environment, which release instinctive behavior. We can imagine the behavior (whether fleeing, fighting, or reproducing) as existing somehow within the animal, waiting only for the appropriate signal, whereupon it is released. The parent bird does not "love" its offspring; it simply follows the dictates of its genes to fill, mindlessly, wide-open mouths of the appropriate color and shape. In fact, certain birds, such as North America's brown-headed cowbird and red-headed duck, as well as the European cuckoo, regularly take advantage of this by laying their eggs in the nests of other species. These foreign young are treated like the parents' own, as long as they possess the appropriate releasers.

Similarly, the releasing of simple pent up physical pressure in the breast is probably more instrumental in stimulating nursing by female mammals than is conscious solicitude for the infants' welfare. Among most lower animals, in fact, parental behavior is sufficiently simple and short-lived to be safely relegated to specific automatic mechanisms, evolved by natural selection. For example, many animals learn the physical characteristics of their young when they give birth, during an extraordinary process of "one trial learning," known as "imprinting." As a result, females of many species learn irrevocably to recognize their young. The small tropical jewelfish, for one, does not instinctively know what her offspring look like; she carries her fertilized eggs in her mouth, incubating them there, and does not set eyes on them until they emerge as small, distinctively colored fry. Then, she becomes imprinted onto them, and will accept only them. If, however, an enterprising biologist substitutes young of another species before the mother jewelfish has ever set eyes on her own, first mouthful of offspring, she will subsequently prefer these foreigners to her own brood.

A mother goat will reject her own kid if she is denied the opportunity of smelling it within a few minutes after birth; a very brief exposure normally fixes the kid's identity in the mother's memory. If the newborn is experimentally removed immediately after birth and replaced by a different animal, the mother will become imprinted onto it, rejecting her natural offspring from then on. These behaviors might seem to indicate poor planning on evolution's part, since such blind instinctive obedience appears to leave much room for error. But actually, malfunctions of the type just described are limited almost entirely to cases of human intervention. When it occurs in nature, imprinting provides a simple, almost foolproof way of assuring parental recognition without overburdening the gene pool with unnecessary information.

Human beings pose a different situation. Since we require an exceptionally long training period, since we are dependent upon our parents for a remarkably long time, and since we behave in such complex ways as to have required intelligent and flexible responses from our ancestors, reliance upon simple releasers or imprinting would clearly be unsatisfactory. In our case evolution needed a particular mechanism, not previously used by animals, for maintaining the necessary close and attentive relationship between parent and offspring. The answer was love.

We encountered this word when discussing another novel human innovation, the intense male-female pair bond. The use of one term for these rather different phenomena is interesting and revealing. The love of two adults for one another is clearly different in kind from the love of parent for child. But our use of the same word for both may suggest an underlying similarity between the two relationships. Both carry strong emotions. So powerful is their hold on us, so pervasive their influence, that we generally find it impossible to step back and view them as *phenomena* that are almost certainly part of our biological makeup and therefore likely to be of some significance to evolution. Both are also heavily invested with cultural significance.

Both parental and sexual love also serve the same ultimate function, the maintenance of a close, long-lasting, well-coordinated relationship between individuals. Both forms of love serve an essential evolutionary purpose. To some extent, both are founded upon a rather banal sounding, but nonetheless significant occurrence: simple physical proximity and the familiarity among individuals that results.

Now admittedly, familiarity itself does not necessarily produce love (under certain circumstances, we are told, it can even "breed contempt"), but it can go a long way. There is increasing evidence from animal studies that simple exposure can produce increased tolerance, and eventually attraction. When a female song sparrow is ready to breed, she cautiously approaches the territorial boundaries of a likely mate. Up until now, he has been busy establishing his own special area, regularly defending it from intruding males who would dearly love to carve away a parcel for themselves, or appropriate the whole thing. The resident male is therefore very aggressive and shows no gallantry toward the female, who after all looks like just another trespasser and is attacked unmercifully. (There is virtually no difference in the appearance of male and female song sparrows.) But instead of either retreating or fighting as an intruding male would, the romantically inclined female simply flutters nearby, returning again and again, just "hanging around" near her attacker. At last, her patience is rewarded as the aggressiveness of the territory owner finally abates. To the observer, it seems that he has gradually gotten "used to" her presence. If they were human beings, we would say that they have fallen in love.

A simple yet fascinating experiment has demonstrated the effect of repeated exposure in producing attraction and even preference in mammals. Laboratory rats were divided into three groups: all were treated equally except that one group was given regular exposure to recordings of several Mozart symphonies; the second group heard the dissonant music of the twentieth-century composer Arnold Schoenberg. The third group was not exposed to either composer, so as to reveal whether inexperienced animals had any initial musical preference, and therefore, whether the preferences of the experimental groups could be attributed to their earlier music-listening experiences. After nearly two months, the rats were tested for their musical preferences. The third group, which had not been exposed to any music, did not show a strong preference, although on balance, they chose Mozart over Schoenberg. (Thereby revealing, one might conclude, the fundamental good judgment of untrammeled biology.) The first group, the one that had been initially exposed to the Mozart, showed an even stronger preference for his music when given a choice. The Schoenberg-exposed group, however, turned out to like Schoenberg. Each of the two experimental groups thus developed a definite preference for whatever it had been exposed to earlier, even when—as in the case of Schoenberg—this stimulus was held in disfavor by inexperienced animals. Familiarity, we must conclude, does not always breed contempt. For most living things, in fact, it produces preference.

It is a long way from musical rats and courting song sparrows to love in human beings, but the analogy seems to hold. Among male-female consort pairs in baboons and human beings, the partners spend long periods of time in each other's presence, often doing "nothing special," simply strengthening their affiliation by their mutual presence. In fact, sexual relations between human beings are often considered degrading if they are just "one night stands," lacking the extensive precopulatory period of increasing association and affection that somehow makes a relationship "meaningful." We can always rationalize this phenomenon by pointing out that getting to know each other enables deeper understanding, and hence, appreciation and ultimately love. But that's just the point.

Parents nearly always love their children. Why? Insofar as evolution is concerned, because children are a major route to biological success. As to the means of achieving this adaptive end, prolonged exposure is probably the key, coupled with the child's possession of the appropriate physical releasers: large head compared to the rest of the body, "cute" little nose and ears, the indescribable appeal of adult traits in miniature, unsteady gait—all of these are characteristics that evoke cuddliness and care-taking in Homo sapiens. We clearly do not succumb to immediate imprinting as does the nanny goat. A parent's first view of its newborn baby, somewhat discolored and possibly with misshapen head due to stresses during birth, often causes more disappointment than devotion. But ugly or cute, boy or girl, scrawny or fat, a child "grows on you" until quite soon it is the object of considerable love and devotion. The child need not even be the biological offspring of the prospective parents, as the great success of adoption has shown. Given the appropriate physical characteristics, evolution has arranged that simple familiarity will generate love, with all its attendant selective advantages.

It is always possible that the ultimate evolutionary significance of male-female and parent-child love relationships is coincidental or at most, tangential to their occurrence. We all accept the notion that human beings have a capacity and in fact, a need for love. "People who need people," we are told, "are the luckiest people in the world." We have a capacity for eating food as well, and indeed, a need for it, but no one considers him or herself especially blessed as a consequence. No one questions the evolutionary utility of food: why should love be different? Thus, loving relationships clearly satisfy a powerful and necessary urge within us, no less real than the adaptive urge to eat when hungry or to scratch when itchy. And just as we can develop pathologies of eating, from obesity to anorexia, we are also subject to pathologies of loving, although most of us do a remarkably good job of balancing our nutrient intake and our loving as well. In the wild state, animals such as goats must be prepared to move quickly right after birth; consequently, they are at some risk of misidentifying their own offspring, since these offspring are so precocious that they are walking and running when just a few hours old. The rigid imprinting between nanny and kid therefore makes good evolutionary sense.

For human beings under prehistoric situations, just as among nontechnologic people today, the chances of a newborn infant being attributed to the wrong mother were probably close to zero. However, in modern maternity hospitals, with sometimes literally dozens of newborns being "processed" daily, mix-ups can occur, and significantly, when they do, we are virtually as helpless as our own neonates.

Other animals show differences in their ability to recognize their offspring, differences that help illuminate our own otherwise puzzling limitations. Consider two species of birds, the rough-winged swallow and the bank swallow. Both dig themselves burrows where they nest in sand quarries and natural clay banks, but whereas pairs of rough-winged swallows nest in isolation from one another, bank swallows nest in dense colonies that may include hundreds of other birds. Ethologist Mike Beecher has been able to show that although bank swallows quickly learn to recognize their own young, and will refuse to feed a stranger who happens to land at their burrow entrance, rough-winged swallows apparently cannot tell one young bird from another. Because they breed in such dense colonies, adult bank swallows during their evolutionary history were probably often faced with the opportunity of caring for young that were not their own biological offspring. Those that did so were selected against relative to those that only invested in their own offspring. By contrast, adult rough-winged swallows normally do not encounter any offspring other than their own, so there is no reason for them to imprint onto the characteristics of their young, nor do they possess a genetic knowledge of them. So they innocently feed anyone they find at home, and sometimes they can be fooled. Natural selection, in a sense, can count on the likelihood that any youngsters in a rough-winged swallow's nest will be the ones that the mother has deposited there herself.

Human beings are probably more like rough-winged swallows than like bank swallows. That is, because of our preceding thousands of generations during which the probability of parents associating with the wrong offspring was very low, we lack innate recognition mechanisms, thereby rendering us susceptible to the maternity ward mix-up. This is an admittedly rare liability, one that is uniquely a result of medical technology and assembly-line maternity procedures. But along with it comes a rather substantial advantage: because we lack such recognition mechanisms, we have the capacity of adopting children not biologically our own, and of developing responses to them that are every bit as profound as those toward our genetic children. In short, we can have "cultural children" instead of, or in addition to, our biological children.

Once again, however, the picture is complicated. Our biologically mediated capacity to accept children that are not biologically our own, combined with the growing (culturally mediated) propensity for parents to divorce, places increasing numbers of children among nonbiological parents. Fortunately, stepfamilies can "work," in large part because we lack automatic rejection mechanisms, and because of the familiarity effect, as well as the fact that we are also conscious creatures who presumably enter into relationships with caution and often, a benevolent desire to make the best of situations. But stepfamilies are also susceptible to the ill effects of biological selfishness, since stepparents and stepchildren are usually intensely aware of their "step" relationship. It may be no coincidence that evil stepmothers and stepfathers are so often the villains of fairy tales: a disproportionate frequency of child abuse and child molestation involves the stepfather, whose involvement with the child comes as a result of his affiliation with the child's mother, not with the child itself. This pattern, although not excusable just because it is congruent with evolutionary expectation, is at least more understandable as a result.

Clearly, not all nonbiologic parents are child molesters or abusers, and stepfamilies can be wonderfully successful. But a glimpse of this increasingly widespread culturally mediated phenomenon through the lens of biological evolution provides a perspective that may be a healthy corrective to the frequent misconception that everything should always go smoothly . . . and then, the guilt and recriminations when it does not.

Both the male-female and parent-child relationships are commonplace situations that are almost universally taken for granted and rarely scrutinized. When considered in the light of evolutionary theory, in fact, they initially appear out-of-place and strange. What had seemed so natural is now puzzling, since the immediate costs are real, but the benefits, diffuse. Parental care seems such an obvious thing to provide, yet when the manifold personal disadvantages are considered, it is suddenly hard to explain. But upon further analysis, these behaviors once again assume a legitimate biological role, and with the added insight, we may well be better prepared to deal with culturally inspired perturbations so that the many faces of love can all make beautiful sense once more. It is said that before one studies Zen, the mountains are merely mountains and the flowers, flowers. To the earnest monk struggling to unravel their deeper meaning, the mountains and flowers are no longer what they had been. But when true enlightenment is finally achieved, the mountains are once again mountains and the flowers are flowers once again.

Around every person there is a circle or group of kindred of which such person is the center, the Ego, from whom the degree of the relationship is reckoned, and to whom the relationship itself returns . . . A formal arrangement of the more immediate blood kindred into lines of descent, with the adoption of some method to distinguish one relative from another, and to express the value of the relationship, would be one of the earliest acts of human intelligence.

—L. H. MORGAN (1871)

PARENTAL behavior would seem to be an obvious example of selfless behavior with a clear evolutionary function. We can also extend the biological explanation to cover such "altruistic" traits as the famous broken-wing display of the killdeer. If a fox approaches the nest of this common little shorebird, the adult will flutter away a short distance, seemingly in distress and feigning a broken wing. As the fox follows, the killdeer flaps away a bit further, seeming always to muster just barely enough strength to escape at the last instant, thus deceiving the fox into thinking that it will be easy prey. By this ruse, the predator is lured some distance from the nest at which point the crafty killdeer, its wing now miraculously mended, flies to safety. Such behavior is not without danger to the killdeer itself. But the benefits of increasing the chances that its offspring will survive have apparently been great enough for broken-wing feigning to have become firmly implanted in the species' biology.

Among many animals, the first one seeing potential danger will immediately give an alarm call, alerting others in the area. In some cases, the calling individual may not have any offspring at the time and thus could not be conferring any survival advantage upon them. In addition, the animal sounding the alarm may actually be at a disadvantage relative to the others, who gain by the alarm caller's apparent altruism. By sounding an alarm the sentry warns others of danger, at the same time making itself more conspicuous and hence, in greater danger than if it had simply kept quiet. It would have been more self-serving to flee silently or hide inconspicuously, leaving its fellows to find out for themselves that danger threatens. So why are alarm calls so common?

This knotty problem—and indeed, the whole paradox of animal altruism—was clarified if not entirely resolved by the British geneticist W. D. Hamilton, who pointed out that reproduction is only a special case of the more general phenomenon of natural selection acting on genes rather than individuals, groups, or species. Parents are especially concerned with their offspring because those offspring are a primary vehicle for the parents' own genes. In the same way, other genetic relatives are also carriers of one's genes; the closer the relationship, the greater the probability that a copy of a particular gene, present in one individual, will be present, by virtue of their common descent, in another. In fact, the preceding sentence can be turned around and made even more accurate: we define closeness of genetic relationship by the probability that copies of genes present in one individual will be present in another.

Hamilton's insight—commonly known as kin selection, or inclusive fitness theory—has emerged as one of the key concepts in sociobiology, the branch of evolutionary biology that is particularly concerned with the study of animal and human social behavior. The probability of shared genes turns out to be crucial evolutionary currency, and a major determinant of social behavior, in human beings as well as other animals.

Many species live in groups containing close relatives and the same explanation, in effect, can account for altruistic behavior within those groups as for parental behavior, provided that some sort of family relation exists between the altruist and the beneficiaries.

Picture a gene "X" that induces its carrier to utter alarm calls, as opposed to a gene "Y" that causes it to remain silent, selfishly. By calling an alarm, an individual carrying gene "X" would save the lives of some of its relatives, thus increasing the gene's frequency. At the same time, by exposing its carrier to higher predation, "X" would tend to decline in abundance. The ultimate balance would depend on three factors: the danger to the altruist, the benefit to the recipient, and the genetic closeness of the two individuals. The closer the relationship, the greater the probability that an alarm-calling gene will be warning itself, and therefore, the higher the risk that will be run until finally, with the very close parent-offspring relationship, risks may be very high indeed and yet still acceptable. On the other hand, natural selection would not generally promote altruistic self-sacrifice on behalf of distant relatives or total strangers.

Anthropologists have long recognized, and long been puzzled by, the fact that human beings the world over tend to organize their social life around systems of kinship. Some form of kin recognition and nepotism is found in every human society. We are obsessed with genes, even those among us who don't know DNA from dynamite. It is quite possible that the entire gamut of human kinship systems, from toleration of an undersirable house guest simply because he's a relative, to nepotism in business and government, is a by-product of this fundamental evolutionary phenomenon. Why do we bother identifying our relatives? Why do we treat them any differently from perfect strangers? On an immediate level, the answer is simple enough: we generally love our relatives, spend time with them, and trust them, at least, more than total strangers. Familiarity, perhaps. On an evolutionary level, such behavior is also in accord with natural selection.

Unlike most of the other examples discussed here, the relation between biological and cultural evolution in the case of human kin selection appears to be mutually supportive rather than conflicting. Insofar as the complex superstructure of kinship systems provides opportunities for appropriately directed biological altruism, cultural frameworks of this kind seem to be adaptive in the evolutionary sense. It is striking to consider that our common behavior toward relatives may have a direct basis in biological evolution. But some problems have arisen, owing largely to the rapid cultural advance of the past 30,000 years. For example, when our ancestors lived in small hunting-gathering bands that probably did not exceed 100 individuals, it is likely that kinship relations linked most of the membership. Selection would thus favor genes for altruistic defense of the group: analogous to gene "X" in our example of alarm calling. Such defense would be especially likely among the males, who are more directly engaged in competition, as well as better adapted for combat.

This could occur even if the defenders did not have offspring, or if immediate benefit to their family was not apparent. It is tempting to fall into the trap of arguing that such behavior is simply a result of "human nature" and that an evolutionary explanation is therefore neither relevant nor necessary. But as in the case of human love, discussed earlier, when we ask why nepotism is part of our behavioral repertoire, we are led to ask what has been the effect, if any, of culture upon this adaptive biological system.

The biological and behavioral system of nepotism was probably quite satisfactory before culture began its very rapid change. With the growth of tribes, communities, city-states, and eventually nations, we have been increasingly forced to interact with nonrelatives and on behalf of those we do not know or especially care about. We are expected to concern ourselves with the survival of perfect strangers. Along with this increased association with nonrelatives has come a concomitant decline in the association of relatives with each other. The common living arrangement of many nontechnologic people is an extended family in which grandparents, uncles, aunts, and cousins often live together, frequently under the same roof. Such arrangements are almost unheard of in modern America, and the consequences of such a culturally mandated deviation from human biology may well be severe, in terms of subtle psychological stress.

Child rearing, for example, is a different endeavor when it falls entirely on the shoulders of a small, nuclear family, as opposed to being a communal responsibility in which other, nonparental relatives are intimately involved. It is a basic principle of engineering that stress is reduced on any single point in proportion as it is spread across a large area. Similarly, the concentration of child rearing on just two adults (or worse yet, just one) seems almost a sure prescription for stress. And older adults, as well as younger nonreproductive ones, may be denied an otherwise satisfying outlet for their basic predispositions when they are denied regular interactions with grandchildren, nieces, or nephews.

Once upon a time, in short, an individual's evolutionary interests were adequately served if she simply insured the well-being of herself and her immediate associates. (As for himself and his immediate associates, this may have been less true, because of the phenomenon of male-male competition, discussed earlier.) But virtually overnight, things have changed. Our technology and economic system have made us dependent upon literally millions of people we will never know, tying our well-being to the fate of a large, rather arbitrary unit, the nation (and increasingly, the world) whereas our fundamental focus remains local: neighbors, family, and self. Cultural evolution has seen to it that "no man is an island," but we may lack the biological vision to appreciate this fact.

Our culturally determined social units have thus suddenly expanded to encompass more than just our relatives, while our biology has been especially primed for protective behavior only toward them and our close associates. Defense of these new larger units and cooperation within them have accordingly become a real problem. The defending individual must somehow be motivated to behave altruistically toward a unit that exceeds anything for which our biological nature has been prepared, and the task therefore falls upon culture to generate support for its own institutions. Human ethical systems are diverse, as expected in a thoroughly cultural phenomenon. They are also very similar, however, in that each seeks to bridge the gap between human biology and human culture: the former seeking selfish benefit, the latter demanding self-restraint at least, and often self-denial or even altruistic self-sacrifice.

On one level, then, we expect that human beings will be altruists; that is, although their actions may endanger or otherwise mortify their bodies, these acts (such as parenthood or nepotism) are in fact selfish on a genetic, and hence, an evolutionary scale. As society views us, however, human beings must appear to be profoundly and dangerously selfish, for the same reason: we tend to look out only for our own personal benefit or that of our relatives. Social psychologist Donald T. Campbell has suggested that there may be a deep quasi-evolutionary wisdom in the various systems of conventional religions and ethical morality that circumscribe the boundaries of acceptable human behavior, since without such restrictions the competitive urges of a normally selfish creature could be terribly destructive to society.

The English philosopher Thomas Hobbes suggested just this more than 300 years ago, when he pointed out that human beings tend naturally to a "warre of each against each," unless restrained by the power of the state, the *Leviathan.* Historically, such restrictions have been seen as necessary and indeed, laudable, although to be sure, without understanding their likely biological basis. In recent years, it has become fashionable to criticize such arrangements as artificial, unduly restrictive, and generally harmful. "Do your own thing," "If it feels good, do it," and other slogans of personal liberation were spawned by a sense of societal oppression, combined—especially during the 1960s—with a sense that established society, with its pollution, its vicious war in Vietnam, its untrammeled materialism, had no right to suppress the individual.

But ironically, those people most likely to support personal rebelliousness of this sort are also most likely to maintain that society itself has certain responsibilities, which go beyond the pursuit of private gain. It may be that an understanding of Homo sapiens' underlying selfishness—enshrined but not necessarily legitimized by the New Right in the 1980s—may once again swing the pendulum toward acceptance of society's rights and obligations vis-à-vis its otherwise unruly members. In the process, however, we had best be alert to the potential for abuse, something that nation-states—whether motivated by ideology of the far right or the far left—seem to find deeply compelling. Certainly, the continuing tension between human inclinations (in large part the product of biological evolution) and society's restrictions (the product of cultural evolution) can be destructive as well as restorative.

When it is widely perceived that individuals are unruly and troublesome, people seem more likely to support social functions such as legal restraints, taxation, police systems, and government structure in general, all aspects of organized society that appear to be fundamentally benevolent. However, there is a danger lurking here. By proclaiming that humanity is fundamentally flawed, riddled with a kind of original sin (whether generated by natural selection or God), we help lay the intellectual groundwork for a whole range of repressive and often despotic social processes. If we assume that human beings are "bad" at heart, then society must be protected against its own members. Following a period of worker discontent and near revolt in East Germany during 1953, the authorities announced that they were "disappointed" with the people, which prompted Bertolt Brecht to suggest that perhaps the government should therefore disband the people and elect another! Brecht, a dedicated Communist and supporter of the rights of society against those of the individual, nonetheless saw the danger of perceiving society to be more valuable than its members. The next step is a fateful one: toward the "national security" state, in which the security of that society, rather than the safety of its component individuals, becomes the only legitimate goal of collective action. Even democratic societies can, and have, walked down this potentially destructive road, especially in the modern era of looming nuclear holocaust. En route, other societies take on the appearance of evil incarnate, thereby justifying attitudes, policies, and the possession of weapons which in their own right would otherwise be seen as totally unacceptable.

Our tendencies to view each other with distrust may also be a direct outgrowth of our tendency to distrust ourselves, and then to project such fear onto others. Writing in *Psychology and Religion* in 1937, Carl Jung put it as follows:

> This terrifying power which nobody and nothing can check is mostly explained as fear of the neighbouring nation, which is supposed to be possessed by a malevolent fiend. Since nobody is capable of recognizing just where and how much he is himself possessed and unconscious, he simply projects his own conditions upon his neighbour, and thus it becomes a sacred duty to have the biggest guns and the most poisonous gas. The worst of it is that he is quite right. All one's neighbours are in the grip of some uncontrollable fear, just like oneself. In lunatic asylums it is a well-known fact that patients are far more dangerous when suffering from fear than when moved by rage or hatred.

Culture can have a useful, modulating effect on human fear and selfishness, leavening our nastier inclinations for the benefit of the group of which we are a part, and upon which we depend. As groups have grown, however, they have arrogated greater destructive power unto themselves, and at the same time have increasingly distanced themselves from the interests of the individuals who comprise them. Environmental degradation and nuclear weapons are cases in point: national leaders now find it expedient to follow policies that could lead to permanent pollution and/or resource shortage, as well as even threaten the destruction of life on earth in order to advance the "national interest." At such a juncture, the various techniques by which culture has overridden our biological proclivities for selfishness cease being benign and become malignant, possibly lethal.

Individuals do not usually give up their autonomy without a struggle. The tendency for self-preservation is strong, like that of reproduction and for about the same reason. In most animals, overcoming the instinct for self-preservation therefore requires potent evolutionary recompense. Respect for the smooth functioning of society is one thing. Permitting it to destroy oneself is another. How, then, does the modern nation-state recruit support for policies that are at best life-threatening and at worst, life-annihilating?

Most social monkeys are capable of organized group defense; in fact, this is one of the main reasons they are social in the first place. Although such defense may consist largely of bluff and show, it can be exceptionally ferocious, with vigorous coordinated attacks in which the enthusiasm of each defender is stimulated and magnified by the behavior of his comrades. The capacity for organized group aggressiveness in these animals may well have a genetic component, and some representation in our own species as well. It is known that human beings in groups, whether organized (an army) or unorganized (a mob), are capable of aggressive activities only rarely perpetrated by solitary individuals. We tend to lose our individuality, our conscience, and our restraints once we become part of a group, and the result can be violent in the extreme. And it seems that we positively yearn to make such an association, to lose ourselves while identifying with a larger collectivity, perhaps because in doing so, we "find ourselves," satisfying primitive adaptive wishes for family and tribal identities in the larger social unit. By exploiting this yearning through speeches, slogans, martial music, and all the trappings of national chauvinism, combined with an intensive program of group indoctrination beginning with the youngest children, societies have been remarkably successful in recruiting willing supporters, especially in times of real or imagined external threats.

Despite the fact that the nation is not, in fact, a biologically meaningful entity, nationalism has been able to achieve a remarkably strong hold on Homo sapiens, apparently because of the strong human tendency to establish social bonds, bonds of the sort that on a personal level are ultimately in our biological self-interest. Hence, nationalistic pseudo-relationships are generated via appeals to fictive kinship (*fraternité*) within the "fatherland" or "motherland." Ironically, in a world of nuclear weapons, the biologically appropriate, pro-life inclination of human beings to establish adaptive relationships with others has been manipulated by and directed toward an artificial, oversized, culturally defined, and increasingly unresponsive political unit—the nation—that threatens human life far more than nourishes it.

Similar conflicts between biology and culture are also apparent in more direct disputes regarding the personal defense of the society: specifically, soldiering. Animals don't behave as if they want to die. And neither do most people. They are unwilling to endanger their lives unless the benefits somehow outweigh the risks. When the risky behavior will result in a generally higher frequency of genes for such behavior, living on in other bodies that we call relatives, this behavior will typically be maintained because evolution favors it, via kin selection. But when the behavior confers only a very diffuse benefit for the genes concerned (the relatives), and when the risks are comparatively high, evolution will select against such tendencies. Defense of a nation, as opposed to a family, can be just such a case, in which we would expect individuals to balk at the more extreme demands of nationalism.

Both historically and biologically, soldiering may have been adaptive for the population as a whole. However, the advantage, if any, to a warrior is often low compared to the risk encountered, especially under conditions of modern warfare. Insofar as natural selection is concerned, individuals typically are motivated by individual benefit, not that of the group; evolution would judge it a bad bet, and select against such tendencies. Furthermore, under conditions of large social units, the individual who "selfishly" stays home and refuses to fight would seem to be in a position to leave more offspring than the "altruist" who marches off to war and may not return.

Primitive appeals to nationalism are used to generate popular support for policy agenda, and to recruit soldiers. But although the nation-state can silence dissent, force acquiescence, and in some cases even generate ardent enthusiasm, it is another thing for the nation-state's military apparatus to be assured a continuing supply of willing cannon fodder, since the preservation of one's own life is, after all, a pretty strong evolutionary imperative. One might expect this conflict between biology and culture to be ultimately resolved in favor of biology. But this is to ignore the importance of culture to the physical survival of our species, and to underestimate its resourcefulness in perpetuating itself. It must be recalled that cultures themselves will tend to develop adaptive characteristics through a process of competition with other cultures analogous to natural selection among strictly biological entities. For example, Napoleon's great battlefield successes were due at least partly to his invention of the *levée en masse*, the first time that the manpower of an entire nation was mobilized for military (or any other) purpose. The other great states of continental Europe learned their lessons well, and before the end of the nineteenth century, Germany in particular had a very efficient military draft. In such cases, however, the characteristic is assumed by the culture as a whole and is not part of the genetic makeup of individuals. Although biological selection operating at the level of groups is inefficient compared to the same process operating at the level of individuals or genes, cultural selection operating at the group level can be very potent indeed.

In his book *The Parable of the Tribes*, theologian Andrew Bard Schmookler examined the problem of power in human social evolution, reasoning as follows: "Imagine a group of tribes living within reach of one another. If all choose the way of peace, then all may live in peace. But what if all but one choose peace . . . ?" The result, as he sees it, is a system driven to ever greater accumulation of power—military, technologic, economic—because of competition among societies. In such cases, individuals are ground under by the desperate competitive machinations of their larger social units.

Forced conscription, as we have seen, is probably society's

simplest and crudest tool for insuring defensive altruism in a nonbiologically motivating situation, when salaries are inadequate to compensate for the risk and other costs of military service. Tax resistance, nonviolent civil disobedience, counterculture life styles, resistance to a military draft: all these and more are thus in the final analysis among the many possible manifestations of the conflict between biology and culture. More generally, resistance to "the system," when it becomes indifferent to life, or worse yet, life denying or life destroying, is often a *cri de coeur* abetted by any number of cultural techniques and prescriptions, but fundamentally a cry from the heart of life itself.

> *If there were only two men in the world, how would they get on? They would help one another, harm one another, flatter one another, slander one another, fight one another, make it up; they could neither live together nor do without one another.*
>
> —VOLTAIRE, *Philosophical Dictionary*

THE HUMAN capacity for group-oriented behavior is heightened by another characteristic of many social animals: rejection of foreigners. Most bees, wasps, and ants, for example, carry a particular odor, specific to their own hive, which permits other hive members to identify them. Foreigners, or residents that have been treated with foreign odors, are generally driven away or killed. Rats live in organized packs whose members are identified and tolerated. Woe to the strange rat who is introduced into the midst of a harmonious pack. The large ground-dwelling monkeys (macaques and baboons), whose behavior may be so suggestive of our own a few million years ago, live in tightly organized groups in which strangers are only rarely tolerated and aggression between groups often occurs. Once again, this is a pattern that makes sense in terms of biological evolution, since members of these relatively closed groups are often genetic relatives as well.

Among many tribal peoples, the word for human being is the name of the tribe: therefore, members of a different tribe are by definition not human beings. It is no coincidence that among many of the head-hunting tribes of the Amazon, killing a fellow tribesman is murder, whereas killing someone else is "hunting." By defining only their friends and relatives as truly human, members of the in-group are free to behave toward members of the out-group in ways that would not be socially acceptable toward fellow tribesmen, or biologically acceptable toward genetic relatives.

Killing of a fellow tribe member is generally prohibited, but killing someone from another tribe may be encouraged. After all, a member of a strange tribe is not a human being. This is not mere sophistry; it is a fundamental fact in the lives of many people, and it speaks eloquently of a world view in which we can identify evolution's oft-unpleasant hand. Kin selection is relevant to this murderous double standard, since a fallen stranger is less likely to be carrying genes in common with his murderer.

In proposing a biological tendency for xenophobia, we are not necessarily predicting 100 percent expression of this trait in all people or all groups: remember that genes define a range of possible expression, rather than rigidly determining a precise characteristic. This seeming equivocation is especially true for behavioral traits that can easily be modified by culture. It should be clear, however, that insofar as xenophobic tendencies exist, they generate troublesome susceptibility to propaganda picturing foreigners as criminal, immoral, not quite human, and certainly untrustworthy.

The human tendency to form groups with insiders and against outsiders is reflected in many aspects of our lives, acting within cultures as well as between them. It begins with children "ganging up" on one another and progresses through exclusive clubs, fraternities, sororities, union locals, service clubs, and political parties. In addition to such purely cultural affiliations that are achieved largely by choice and by personal effort, there are cultural units into which one is born and for which choice, although possible, is experienced only rarely. Religion, ethnic group, and nationality would be prime examples. Beyond this, conspicuous physical differences with a genetic basis provide a convenient rallying point for human discriminatory tendencies. Racial differences in skin color are a prominent case. And when such differences do not exist, we manufacture them, via clothing style, language, accent, secret passwords, astrological sign, or other totemic association.

We human beings have a notable inclination to exclude individuals who are conspicuously different from us in any way. Such behavior in its relatively primitive untrammeled form is probably biologically adaptive, for the following reason: among most animals, disease is a prominent cause of mortality, probably more important than is generally realized. Since many diseases can be transmitted by infected individuals, it would be advantageous if diseased animals were somehow prevented from associating closely with the healthy ones. Thus, in many animal societies, diseased or disfigured individuals are often mercilessly hounded and excluded from the group. As a general rule, those that are different get ostracized.

Unfortunately, "differentness" in human beings is much more often a function of opportunities, ideas, and inclinations (i.e., culture) than a biological condition. Loners, eccentrics, men with long hair and beards, people who go barefoot or wear beads, women without bras, "kooks" of all kinds become the subjects of society's antagonism. Only firm cultural insistence upon tolerance, motivated by recognition that ultimately society is best served by maintaining freedom and diversity, can save us from the stifling homogenization that would result from such evolutionarily generated xenophobia run wild.

At evolution's behest, we thus tend to defend and protect our offspring, favor our relatives, identify with groups, respond to mob psychology with a propensity for violence, and distrust outsiders and anyone who is different. The interaction of our biological and cultural heritage leaves us enmeshed in a complex mosaic of the bestial and the beautiful, of problems and possibilities.

The biologically evolved human capacity for benevolence and cooperation is not limited to patterns of genetic relatedness or immediate self-interest. There is yet another mechanism by which natural selection can get a handle on altruism. Known as "reciprocal altruism," or more simply, "reciprocity," it suggests how apparent altruism could evolve between individuals that are completely unrelated; indeed, even between members of different species.

The crucial requirement is that the beneficiary eventually reciprocates the favor, so that the altruist ends up ahead in the long run. Once again, as with kin selection, biological evolution can select for altruism so long as it is not really altruism at all, but rather, selfishness. For example, among the inhabitants of coral reefs are small, brightly colored "cleaner fish," notably wrasses of several different species. These little iridescent animals enter the gills and even the mouths of larger fish, including moray eels, removing various external parasites. The animals being cleaned profit from the service, and the cleaner fish get a meal. In a sense, the former animals are being altruistic, in not eating the cleaner when they could easily do so, just as the cleaner is being altruistic in not taking a bite out of its host. So each participant gives a little, and in the long run, both are better off.

Now, imagine a prehuman Australopithecine who has just killed an antelope. After stuffing himself and allowing his mate and his relatives to eat their fill, there may be little cost to sharing the remainder with other members of the group, especially if by doing so, the fellow doing the sharing makes it more likely that he will benefit from someone else's largesse next time around. If the cost of the altruistic act is not too great, and the opportunity for reciprocation is sufficiently high, it would not matter whether the participants are related or not (although, in fact, it would help).

Human beings are quite sensitive to reciprocation, typically responding with what has been called "moralistic aggression" when someone fails to return a favor. It may even be that without this capacity, reciprocal altruism would not evolve. This is because reciprocal systems are highly susceptible to being disrupted by cheaters, individuals who accept altruism from others but then refrain from paying them back later.* Such noncooperators enhance their fitness by finding it more blessed to receive than to give, thereby making chumps out of the altruists, who suffer a reduction in their fitness by giving without getting.

As a result, human reciprocal systems are best developed among individuals who know each other well and are likely to interact frequently in the future. We are more likely to trust a neighbor than a stranger, knowing that she is more likely to return a favor. In many cases, bonds of genetic relatedness are overlain with bonds of reciprocity, but among human beings at least, either one or the other can be sufficient to produce cooperation that may ultimately be beneficial to both. Conflicts arise under two circumstances: when the expected reciprocity is not forthcoming, and when suspicions of nonreciprocity interfere with cooperation that might otherwise be helpful or even necessary for the success of both parties. In the latter case in particular, such conflicts also involve biology and culture.

This problem has been recognized by mathematicians, psychologists, and political scientists for some time, and is called the Prisoner's Dilemma. The dilemma is as follows: imagine a simplified situation in which two individuals are interacting, each having the option of doing one of two things, either cooperate (altruistically) or cheat (selfishly). The payoff that each receives depends not only on what he or she does, but also on what the other one does, and yet each is forced to decide independently, not knowing whether the partner will be altruistic or selfish. We can represent the situation as follows, with the words inside the boxes indicating the payoffs received by individual 1:

*There is at least one species of fish, the saber-toothed blenny, which resembles the cleaner wrasse in both appearance and behavior, except that when the larger fish open their gill covers to be cleaned, the blenny takes a quick bite, and then darts away!

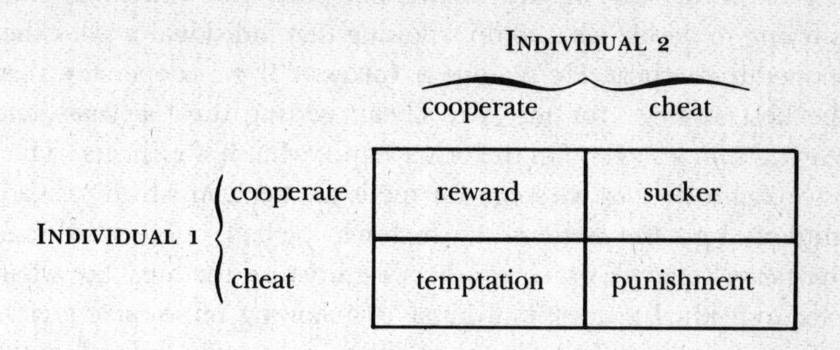

		Individual 2: cooperate	Individual 2: cheat
Individual 1	cooperate	reward	sucker
	cheat	temptation	punishment

(Of course, a mirror image payoff would apply to individual 2.)

Cooperating altruistically brings a *reward* so long as the other individual also cooperates; cheating selfishly and refusing to cooperate brings a *punishment,* so long as the other individual is also selfish; cooperating altruistically when the other individual is selfish makes one a *sucker*; and finally, being selfish and refusing to cooperate yields the *temptation* to cheat if the other individual is cooperative (and also gets suckered in turn). Depending on the payoff values, there may be no dilemma at all. For example, if the payoff for mutual cooperation is high enough and the temptation to cheat is low enough, everyone should happily cooperate. Other patterns can also be analyzed. The most interesting one, however, is when temptation is high, the cost of being suckered is very great, and the reward of cooperation and the punishment for cheating are intermediate, with the former higher than the latter. Under such conditions, the participants are prisoners of a dilemma.

Here it is. Individual 1 would be best off if he cheated and individual 2 cooperated; he would then reap the temptation to cheat. However, individual 2 is in the same position. The two of them would be best off if they both cooperated, since the reward for cooperation is higher than the punishment for mutual cheating. But each is prevented from cooperating by fear that if he did so, the other might cheat, leaving the would-be altruist a sucker.

Look at it this way, again from the perspective of individual 1: he is trying to decide what to do, knowing that individual 2 will either cooperate or cheat. He reasons as follows. "If #2 cooperates, then the best strategy for me is to cheat, getting the *temptation* and leaving him a *sucker*. On the other hand, what if #2 cheats? Then once again the best strategy for me is to cheat, in which case admittedly I get the *punishment*, which may be bad, but less bad than the *sucker's* payoff which is otherwise in store for me." So, whatever individual 2 does, individual 1—following impeccable logic—is forced to cheat. Individual 2, following the same logic, does the same thing. The outcome of the Prisoner's Dilemma is that both individuals find themselves cheating, and receiving the *punishment* whereas if they had both cooperated and behaved altruistically, then both would have received the *reward*. Each one, fearing that the other will cheat and that he will be made a sucker, is forced to be a cheater himself. And so both wind up worse off than they would otherwise be.

"The reasonable man adapts himself to the world," wrote George Bernard Shaw in *Man and Superman*. "The unreasonable one persists in trying to adapt the world to himself. Therefore all progress depends on the unreasonable man." When two reasonable men meet, neither attempts to adapt the other to himself; that is, both cooperate and both receive a high payoff, the reward of mutual cooperation. But when an unreasonable man meets a reasonable one, the unreasonable one persists in trying to adapt the other to himself—that is, he cheats. In the short term, the result may be a kind of progress: the unreasonable man does better than the reasonable man because the unreasonable man receives the temptation for cheating while the reasonable one gets the sucker's payoff. But over time, the system retrogresses rather than progresses. Reasonable people become unreasonable, or else they disappear and are replaced by the unreasonable, who are more resilient because they do not get suckered. The result of such a system, therefore, is a population of all unreasonable people, none of whom cooperate. Admittedly, none of them get suckered, so in a sense their behavior is not entirely unreasonable, but they don't do any better than the relatively poor and punishing payoff of mutual cheating. Just as bad money drives out good, unreasonable or cheating behavior in a Prisoner's Dilemma tends to drive out reasonable or cooperating behavior. (The dilemma, once again, is that all participants would come out ahead if they could only figure out some way of being mutually reasonable, and cooperating.)

Human beings, even the illogical and nonmathematical among us, have probably been shaped by interactions of the Prisoner's Dilemma type. This is not to suggest that we necessarily run through the sometimes tortuous calculations or that our ancestors did. Rather, natural selection did the analysis and as a result, we tend to shy away from cooperating if there is too great a chance that we will end up being suckered. We therefore tend to be suspicious, especially when dealing with strangers who might cheat us. Moreover, we often tend to view the world as a Prisoner's Dilemma, even when it isn't.

Fortunately, however, there are many ways out of the Prisoner's Dilemma. As political scientist Robert Axelrod has described in his important book *The Evolution of Cooperation*, one of the most effective such routes involves "Tit for Tat," the exchange of small reciprocations each of which is relatively cost-free, so that a pattern of mutual trust and mutual benefit result. (It is also necessary, it seems, to retain the option of responding with a "cheat" if the partner/opponent does so.) Another way out is to recognize that mutually advantageous solutions can be available regardless of what the other side does, and that systems that under primitive, biological conditions were Prisoner's Dilemmas may yield different outcomes in modern times. For example, in cases such as overpopulation, resource depletion, criminal justice, or nuclear war, the worst outcome is not being suckered while the other side cheats, but rather, the mutual *punishment* for mutual cheating. Because of relatively recent cultural developments, in a variety of areas, the payoffs for cooperation have increased, along with the costs of cheating. Heedless of these charges, however, the primitive bio-psychology of Prisoner's Dilemma still holds sway.

As the world grows smaller and even strangers become reciprocators if not relatives, the tendency to see the world as a Prisoner's Dilemma becomes itself a dilemma, a dilemma that originates in the discordance between our biology and our culture.

CHAPTER SEVEN

AGGRESSION, KILLING, AND WAR

The Arts of Death and the Hearts of Men

> *I have examined Man's wonderful inventions and I tell you that in the arts of life Man invents nothing; but in the arts of death he outdoes Nature herself, and produces by chemistry and machinery all the slaughter of plague, pestilence and famine . . . In the arts of peace, Man is a bungler . . . His heart is in his weapons.*
>
> —George Bernard Shaw, *Man and Superman*

It is difficult, perhaps impossible, to remain coolly cerebral when discussing aggression, killing, and war, particularly in the closing decades of the war-torn, holocaust-haunted twentieth century. The human mind, capable of marvelous creativity, keen insight, and soaring grandeur, is also capable of the most notorious excesses of ruthlessness, barbarity, and unmitigated horror. It can be the best of things and the worst of things, and never worse than when it hurts and kills. It is also frightening and confusing, not only in what it can do, but even in our uncertainty about what it *is.* Listen to British historian Hugh Trevor-Roper, describing the mind of Adolf Hitler (in his introduction to *Hitler's Secret Conversations, 1914–1944*):

> A terrible phenomenon, imposing indeed in its granite harshness and yet infinitely squalid in its miscellaneous

> cumber—like some huge barbarian monolith, the expression of giant strength and savage genius, surrounded by a festering heap of refuse—old tins and dead vermin, ashes and eggshells and ordure—the intellectual detritus of centuries.

In this chapter, we shall examine some of this detritus which makes up the human species, and which has accumulated not over centuries but rather, millennia. Are we really bunglers in the arts of peace? Do we really wear our hearts on our military arms? And if so, why?

People love to speculate about whether human beings possess an "aggressive instinct." Two rather distinct sides have emerged, one group convinced that we have a genetically mandated tendency to kill our fellows, and the other equally persuaded that human aggressiveness is simply a function of our environments and in no way related to the DNA of our animal past. Most likely, both are wrong. To understand this important controversy, which has generated considerable aggression in itself, it would be helpful to review some history.

Ethology, the science of animal behavior, is a relatively new discipline, owing much of its development to a union of old-time natural history with modern biology beginning largely in the 1930s. It is similar in some ways to the psychological study of comparative behavior, but with many differences in orientation, emphasis, and technique. These differences have produced rather divergent outlooks. For one thing, ethology is still largely a European specialty (although a vigorous following has developed in the United States), whereas comparative psychology is basically an indigenous American movement. Ethologists are biologists, primarily concerned with the study of behavior in an evolutionary context, and they tend to study animals for their own sakes.

By contrast, comparative psychologists are ultimately concerned with human behavior. When they study animals, it is with an eye toward our own species and with the expectation that by studying behavior occurring in both animals and Homo sapiens, they will gain insight into ourselves. Because of their emphasis on the animals for their own sakes, ethologists tend to conduct their studies in natural habitats, where the adaptive value of animal behavior will be more readily appreciated. On the other hand, on account of their emphasis upon experimental manipulation and control, most research by comparative psychologists is conducted in the laboratory.

Even the animals studied are different. Ethologists tend to investigate a wide variety of animals, in order to gain perspective on the diverse results of evolution. They concentrate on birds, fish, and insects. By contrast, comparative psychologists work largely with mammals, which bear greater resemblance to human beings. They have concentrated almost exclusively on studies of the white laboratory rat, an animal whose similarities to Homo sapiens can at least be debated. In fact, the comparative psychologists' preoccupation with white rats has been so great that a leading American behaviorist, Frank Beach, wrote an article in a prominent psychology journal urging his colleagues to widen their range of study animals. In his paper, a cartoon showed a reversal of the Pied Piper myth: A horde of lab-coated scientists were eagerly following a large white rat into the river outside of Hamelin Town.

Ethologists have specialized in studies of species-typical behaviors that are often unique to each species and that generally have a genetic basis. Comparative psychologists have concentrated on the more modifiable behaviors that are characteristic of ourselves. Chief among these, of course, is learning.

Given these underlying differences, it is not surprising that ethologists generally attribute a stronger genetic component to behavior than do psychologists, a difference in outlook that extends to the question of human aggressiveness as well. Thus, the advocates of a hereditary tendency for human aggressiveness have been led by ethologists such as Konrad Lorenz, Niko Tinbergen, Karl von Frisch, and Irenaus Eibl-Eibesfeldt. Within the past few decades, the American playwright Robert Ardrey and the British biologist Desmond Morris have popularized the ethological viewpoint in a series of provocative books. The opposing view, that aggressiveness is largely a result of specific environmental factors ("learning" in the broadest sense) and that peaceful societies are attainable with appropriate manipulation of the environment, has been urged by the American psychologist John Paul Scott and the anthropologist Ashley Montagu, among others. Psychiatrists tend to fall on both sides of the argument. Given their primary orientation toward human beings rather than animals, they have an understandable bias toward modifiable traits, and hence, those due to culture and early experience. On the other hand, the Freudian tradition of unconscious factors, as well as a sensitivity to genetics and biological underpinnings tends to make biological evolution also acceptable to psychiatry as a fount of human behavior.

We are impatient creatures wanting quick answers, especially to such a momentous question as the origin of human aggressiveness. But unfortunately, the best answer at this time is that we simply don't know which viewpoint is correct. It seems most likely that the truth lies somewhere in between. There is good evidence that many animals have a built-in "need" to discharge aggressiveness and that if an appropriate outlet is not available, they may injure their mates or even themselves. Animals frequently indulge in elaborate "appeasement" behaviors to defuse the aggressiveness of another, often a prospective mate. Aggressiveness, particularly between males, is frequently elicited automatically by the simple appearance of particular releasers, like the black mustache of a flicker, mentioned earlier. But many psychologists have also presented evidence for the role of experience in determining aggressiveness.

Scott was able to produce aggressive mice who were sure winners in any bout with an opponent, even a larger one, by previously exposing his candidate to a graded series of fights which were all "fixed" to insure its victory. The mouse that has fought and won in the past is much more likely to fight and win in the future. The implications are that animals learn to fight by fighting, just as they learn to win by winning. Further results also indicate that they learn *not* to fight simply by not fighting. This might suggest that among our own species, strict prohibitions against aggression would ultimately result in a nonaggressive temperament. In addition, Scott's research indicates that fighting is often a result of breakdown in the prevailing social system. One effect of social organization in many species thus appears to be the maintenance of order. Accordingly, by avoiding such breakdown in our own societies, we might help keep the peace.

Other suggestions for environmental control of aggression have also been made. The influential Yale psychiatrist J. Dollard and colleagues proposed long ago that frustration is a major cause of human aggression. Raising children in a frustration-free environment would thus appear to be a logical way to avoid unwanted aggression. (On the other hand, as Konrad Lorenz has pointed out, "nonfrustration" children frequently prove to be intolerable brats, whatever their level of aggression.) It has also been observed that animals often fight in direct response to painful stimuli. If two rats are kept together in a cage outfitted with an electrified grid, they will begin fighting with each other in response to a shock. This so-called "reflexive fighting" provides further undeniable evidence for the role of experience in influencing aggression. Indeed, all these independent hypotheses mesh nicely together, suggesting to some psychologists that a well-organized human society that eliminated frustration, pain, and the opportunities for fighting among children would have eliminated fighting and war among adults as well.

On the other hand, fighting in response to pain is also adaptive, and may well have been strongly selected by evolution. An animal experiencing an attack and the pain that comes with it might be well advised to respond by fighting. The electrified rat may "think" that the other rat is somehow responsible for the pain it has just felt, whereupon it responds—appropriately—by fighting. This argument does not rule out the potential significance of pain (or similarly, frustration). However, it does introduce an important new wrinkle; namely, the adaptive value of animal aggression, whether self-generated or in response to experience.

This was the major point of Lorenz's provocative book *On Aggression.* It is significant that the original German title was *Das sogenannte Böse*—literally, "the so-called evil"—implying that although violent human aggression may be a great evil, aggression itself is a basically adaptive characteristic, and one that evolution has fostered in animals, and possibly in human beings as well. In addition to facilitating simple defense, the capacity for aggression enables animals to assure adequate room between themselves and competitors. Individuals who are sufficiently aggressive are likely to be more fit than those who are wimps. Lorenz in particular has also emphasized that the very important male-female pair bond is supported in many species through the sublimation and redirection of each partner's aggressiveness toward the other, as well as the sharing of aggressiveness directed at outsiders. Given these advantages, the "so-called evil" appears increasingly desirable. It seems reasonable to suppose, therefore, that evolution may have somehow engineered it into the genetic makeup of human beings.

Anthropologists are fond of pointing triumphantly to a few basically nonaggressive human societies, such as the African pygmies, as "proof" that our species lacks a genetically mediated tendency for interpersonal violence. But again, the presence of apparent exceptions does not conclusively prove anything, since genes produce potentiality, not certainty. It is of some significance, moreover, that nonaggressive human societies exhibit great gusto in such other behaviors as eating, drinking, playing, laughing, and sex. This might actually support Lorenz's original idea (now nearly five decades old) that aggression builds up as a spontaneous drive, leading to the proposal that humanity's aggressiveness could be reduced by redirecting it, thereby encouraging more acceptable outlets for our pent-up energies. Furthermore, many seemingly nonaggressive people such as the Kung Bushmen, Australian aborigines, and certain Eskimo groups, have been forced into their peaceful stance after being defeated by other, more aggressive people. The salient fact remains that we are an overwhelmingly aggressive species, and this remains true whether our aggressiveness is generated out of whole cloth from within ourselves, or as a response to personal background, social situations, or frustration.

Lorenz and others have proposed more competitive athletics, on both the participant and spectator levels. Healthy competition—local, national, and international—could also be encouraged in such potentially beneficial fields as space and undersea exploration, socio-economic development, medical research, pure and applied science, and the arts. There is, in fact, some evidence that simply watching an aggressive act serves to decrease the antagonism present in an observer. In experiments, a group of subjects were insulted and thus made angry and aggressive: their blood pressure and pulse rate increased perceptibly. When they were subsequently shown films of boxing matches, automobile accidents, or other violent events, their pulse and blood pressure quickly returned to normal. It therefore appears that simple vicarious experiencing of aggressive release tends to diminish pent-up aggression. Most advocates of environmental control of human aggressiveness would disagree with these proposals, calling instead for the exact opposite: minimizing the opportunities for expressing aggression and violence, rather than encouraging its release even if in a "harmless" manner. They also have their studies to point to: research showing that exposure to explicit violence, especially on television, may well stimulate violence. Whereas the witness to real-life violence is often revolted and appalled, the witness to artificial violence, particularly if repeatedly experienced, seems to develop different norms for acceptable conduct as a result, as well as inaccurate perceptions regarding the actual consequences of violence.

It seems clear that too much aggressiveness is disadvantageous, and just as likely to be selected against among the ancestors of human beings as among any other living things. Since there are possible costs associated with aggressiveness (notably, the danger of being injured or killed) as well as potential benefits, it seems undeniable that being too aggressive is no less maladaptive than being insufficiently so. Beyond the danger of self-injury, hyper-aggressive individuals might harm their mates or relatives, waste time and energy that could be better spent on self-maintenance, and even lead to the phenomenon of aggressive neglect, in which offspring receive insufficient care because the parents are preoccupied with their own threatening, posturing, or fighting.

The "innate" level of human aggressiveness might ultimately be illuminated by the approach of sociobiology, which emphasizes the use of evolutionary biology, and especially the concept of maximum fitness—reproductive success of genes—to interpret and predict behavior. Rather than consider human beings to be either innately aggressive or innately nonaggressive, a sociobiological view suggests that we have been selected to behave aggressively under certain conditions and nonaggressively under others, depending on the consequences of such aggressiveness or nonaggressiveness for our evolutionary success. In short, aggressive behavior that is adaptive under one condition may be maladaptive under others; aggressive behavior that is adaptive for one individual (say, an adult male) may be maladaptive for another (say, a juvenile female). Like a chemist carefully measuring the right amount of an acid necessary to neutralize an explosive alkali, we can expect that in most cases, living things will titrate their high-risk behavior with some precision. Continuing the analogy, we can expect this to be done with much less precision when the substances involved have been newly created, so that their effects are not well known. Natural selection typically assures that living things "know" the reaction ranges of their behavior; but natural selection has not had much opportunity to act on modern weaponry.

In recent years, concern about human aggression run wild has been heightened by our development of nuclear weaponry and the capacity for self-annihilation. By the massive elaboration of our potential for killing, cultural evolution has rendered a possibly adaptive trait highly dangerous. Consider the progression of the "cides": from suicide, the killing of oneself, we can expand to homicide, the killing of someone else. Although certain biblical injunctions called for the next step, genocide—the killing of an entire people—it is only in more recent times that such a potential has been readily available and efficiently acted upon. Shortly afterward, we coined the term ecocide, for the killing of an entire ecosystem; and now, with nuclear weapons, we can look ahead to the latest and most challenging prospect of all, omnicide.

Animals do not view their own aggression as good or bad; it is simply part of their lives, like sleeping, eating, cheating, cooperating, or copulating. Human aggression, on the other hand, readily lends itself to ethical judgment, most often to condemnation. We must therefore concern ourselves both with the situations causing the actual appearance of aggression, and with its results, the ways in which human aggression is expressed. As is so often the case—and contrary to much popular opinion—the problem derives not so much from the instincts we *have*, but rather, from those we *lack*. And both the elicitation of human aggression and its actual expression are strongly influenced by the conflict between culture and biology.

Human beings may possess an instinct for spontaneous aggressive behavior, although it seems unlikely. Conceivably, all human violence results from particular environmental factors such as poor rearing conditions, frustration, social disorganization, personal neuroses or psychoses, etc. But even if human aggression is not generated *de novo* by our genetic constitution, the capacity for such behavior must ultimately derive from our genetic makeup, the results of biological evolution.

For an analogy, consider behavior for which genetic factors are unquestionably responsible: the capacity for complex learning. With sufficient training, a human being can learn to solve difficult problems in differential calculus, whereas even the most intelligent chimpanzee cannot. The distinction between their abilities is very great, almost entirely because of differences in their genetic makeup. Admittedly, an untutored human being cannot do calculus, but even an intensively coached chimp is utterly hopeless in this regard whereas any moderately intelligent Homo sapiens can learn the technique. This is not to say that we possess genes that specifically control the solution of calculus problems, while the chimpanzee does not. Rather, the human genetic constitution includes the capacity for elaborate symbolic and abstract thought whereas the chimp's does not. The fact that African bushmen do not regularly solve calculus problems does not mean that they are incapable of doing so. Their normal environment simply does not provide the appropriate circumstances. At minimum, then, human aggression must similarly be based on a *capacity* for aggressive behavior, which emerges under the "appropriate" circumstances.

Granted, then, that human aggression must derive at least indirectly from our biological makeup, one aspect of the aggression problem can be seen as springing from the interaction of our culture with this biology, since most human cultures provide situations that actually stimulate human aggression. Given our biological capacity (if not need) for aggression, many human cultures may even make its expression more likely than it would be if we lived with fewer cultural complexities. All this, in social systems that may avowedly seek to discourage aggression.

Violent incidents are rare among animal societies, in part because the order of social precedence is generally well known and adhered to, based on a biological capacity for establishing and maintaining such relationships. A hen in a flock, a cow in a herd, and a baboon in a troop all know their position relative to their colleagues, and fights are generally avoided—or if unavoidable, minimized—by the lower-ranking animal's deference to the higher. The system works so well that even such seemingly mild indications of dominance as threat or displacement of the subordinate by the dominant are rarely seen, because the former judiciously avoids confrontations with his or her superiors.

We apparently lack a well-developed biological capacity for social harmony attained in this manner; the despotism of such relationships among animals often seems viscerally unpleasant to us. We have thus substituted cultural rules for biological imperatives, requiring courts, jails, and police to accomplish what most animals have unconsciously achieved under evolution's guidance. Laws and police are uniquely human institutions; the baboon's policeman is his own biology.

This is not to claim that animal social systems always work perfectly or that tumult and sometimes even violence are unknown; rather, there is a predictable and consistent pattern to the ordering of animal social life. When disruption occurs, it is typically to the ultimate benefit of the disrupter. A simple experiment showed that among the mountain-dwelling Gelada baboons of Ethiopia, for example, individuals are remarkably calculating about whether or not to "make trouble," thereby showing considerable enlightened self-interest. Thus subordinate, mid-ranking, and high-ranking males were each exposed separately to different male-female pairs. High-ranking males typically intervened and tried to win the female for themselves; given their high social status, they were generally successful. Mid- and low-ranking males, on the other hand, typically did not intervene, so long as the behavior of the female in the male-female pair showed that she was closely affiliated with her male. In such a situation, a would-be sexual usurper would not only have to deal with the outraged "husband," but also with an uncooperative "wife," whose loyalty might be difficult to achieve. But when the female seemed bored, disaffected, or otherwise inattentive to her male, then even a mid- or low-ranking male would try to win her away.

Once established, social hierarchies among animals may be very long lasting. The "alpha" or number one animal often remains the uncontested dominant long after his physical powers have waned. He rules by reputation. Human beings similarly establish hierarchies based on reputation, but they are considerably less inviolate; indeed, they are constantly being tested. The result is almost continual emotional stress and occasional physical violence. Even if we did have a biological tendency to respect a social order once it is established, we would rarely have the opportunity to do so because of our advanced cultural systems: we encounter complete strangers all the time, people whose relation to ourselves cannot readily be determined.

Because of the efficiency with which animal social organization generally keeps the peace under natural conditions, students of animal behavior often have difficulty identifying the relative social ranking of particular individuals. A good technique for solving this problem is to introduce small amounts of some desirable commodity such as food, and watch them fight over it, or (more likely) see who defers to whom. With the savannah-dwelling baboons this is especially effective, since these animals are normally spread over a wide area, eating things like grass or roots that are evenly distributed over their range. Aggression may be induced among such animals by providing a situation of extreme competition for some peanuts or an apple.

Woodchucks are rather surly, independent, and almost solitary creatures, which become sociable only briefly, at mating. Most of the time, they keep away from each other, eating grasses and seeds. The males will sometimes fight, especially during the breeding season, but most woodchucks simply avoid one another. After all, their food is widely distributed, over a broad area, and usually there is nothing much to fight about. This is changed, however, when someone plants a small garden: suddenly, many animals are attracted to a limited area, where they rub shoulders in close proximity, and get mad at each other. In addition, they now have something to fight about: dozens of succulent pea plants, for example, in just ten feet of carefully cultivated row. Like the Greek gods and goddesses who began bickering among themselves when Eris, the goddess of discord, presented them with a golden apple inscribed "for the fairest," woodchucks are more likely to fight, wound, and kill each other when they have something clear-cut to fight over.

According to Greek mythology, the exploits of Eris gave rise ultimately to the Trojan War. In real life, comparable situations have persistently generated discord, in abundance. Our culture today places us in direct and continuing competition with each other for a limited amount of an easily identifiable resource: money, as well as social status. These are culturally mediated pressures that a small band of hunter-gatherer Australopithecines would never have experienced. But it is not necessary to look only at technological civilization to see examples of culturally inspired aggression. The struggle for simple physical possessions is a frequent cause of human violence, and the possession of objects of one sort or another is a universal human trait. Thus, the "crime" of stealing, with its frequent companion—violence—could not occur unless there was something to be stolen. Ownership of things external to our bodies is well developed in human beings, and is a direct outgrowth of our use of tools. But it is not unknown in some animals as well, and when ownership occurs, it too is often a source of conflict.

Many hunting animals fight to defend their kill from others, particularly from scavengers such as hyenas or jackals who try to steal it. Among birds, the Antarctic skua and tropical frigate bird are specialized air pirates that regularly get their meals by stealing the catch of such expert fishermen as gulls and cormorants. The gulls, in turn, commonly steal nest material and eggs from other unwary gulls. Male scorpion flies hunt for small prey items, which they present to females as part of courtship; sometimes, a male without such a prey item alights next to another that has one. Acting like a courting female, the empty-handed male attempts to steal the prize, which is normally passed to the female during mating. When he discovers that he has been tricked, the "wealthy" male tries to take back his possession, and a tug-of-war often ensues, sometimes leading to a brief fight as well.

Squabbles may break out over a prized food item among the monkeys and apes, but the disputed object is quickly consumed, and with it, the passions of the dispute. Human beings are unique among primates, however, in that we experience prolonged material ownership. This custom creates long-lasting grudges, and persistent interpersonal violence.

Most primates eat fruits, vegetables, or small invertebrates. In such cases, the food items are usually small and equally abundant over a wide area, thus making competitive struggles uneconomical. The higher monkeys and apes such as baboons, gorillas, and chimpanzees will eat medium-sized animals such as young gazelles and other monkeys when they can get them. The meat of larger animals is rich in protein and highly prized when available. However, a well-established social system generally determines priority at the kill, preventing undue aggression even then.

Among our partially carnivorous ancestors, it is likely that occasions for fighting over a kill were considerably more frequent than they are now among the monkeys and apes. It is therefore possible that we evolved a genetically influenced capacity for establishing and maintaining peacekeeping hierarchies, and that we still carry a remnant of this capacity. Thus, many of our cultural institutions, including systems of royalty, church hierarchies, armed forces ranks, and the various personnel levels on the corporate and governmental ladder may satisfy a tendency for such graded social organization. But the specifics in each case are clearly determined by cultural factors, not genetics. Our social institutions, even at their most despotic, lack the mousetrap guarantee of many of the rigidly instinctive behaviors shown by some of our animal cousins. If such systems were more firmly rooted in our biology, then the various obnoxious devices of tyrannical governments the world over would be unnecessary. (Or alternatively, they would seem less obnoxious!) Torture chambers, secret police, revolution, and counterrevolution all attest to the frequent discordance between our cultures and our biology. It is unclear precisely why human beings have experienced such a relaxation in genetic influence on their social organization. Presumably such flexibility helped us inhabit a wide range of environments, and deal with a wide range of circumstances even when we only occupied the African savannah. In any event, the pattern is consistent with the evolutionary decline in biological control and concomitant increase in cultural control of our behavior generally.

Regardless of whether we possessed, or still possess, a biological capacity for avoiding aggressive competition over food or other items, our culture has vastly increased the opportunities for similar aggression. It has been proposed that instead of Homo sapiens, we should have named ourselves *Homo faber*—Man, the maker—in recognition of our propensity to build and to use tools. No animal constructs as many different things as we do, and none values them so highly. Although a male stickleback fish will vigorously defend his territory, just as a female gull will defend her nest, no animal invests its possessions with the time and labor that human beings commonly do. To expend effort on something is to increase its value, and we put great labor into our things, valuing them very highly indeed, enough to fight and kill over them.

More than twenty years ago, in his popular book *The Territorial Imperative,* Robert Ardrey reviewed some of the ethologists' literature on territorial behavior in animals and used this as "proof" that we are territorial as well. He was both damned and applauded for these efforts. Basically, his arguments are too facile; it simply does not follow, for example, that because certain dragonflies are territorial, so are we. What occurs in animals need not necessarily occur among us.

True, human beings often demonstrate rigid conventions in their social use of space, but these patterns are culturally determined, varying greatly from one society to the next. The anthropologist Edward Hall has shown that significant differences exist, for example, between the conversational distances maintained by Arabs and Americans. Thus, the Syrian or Egyptian characteristically puts his face very close to the person with whom he or she is speaking. A friendly conversation requires that each participant be bathed in the odors and even the body heat of the other. By contrast, Americans and Western Europeans maintain greater interpersonal distance. These varying cultural styles can produce serious misunderstandings at international gatherings such as the United Nations, where persons from "short distance" countries (most of Latin America and many Mediterranean nations may be included here) sometimes appear pushy and overly forward to their "long distance" counterparts. On the other hand, Americans and others influenced by the social conventions of Western Europe may appear cold and indifferent to people accustomed to closer interactions.

Similar cultural differences occur with regard to living space, trespassing, and concepts of privacy. Far from being genetically tied to a territorial imperative, we appear to be susceptible to whatever mores prevail in our particular society.

It has also been claimed that our supposed territorial nature is partly responsible for our aggressiveness. This is doubly unlikely, since one consequence of territorial behavior—as in the case of social hierarchies—is a *reduction* in aggressive incidents. Among territorial species, each individual is necessarily aware of the established rights of his or her neighbors and generally respects them. The exceptions often occur early in the breeding season or at other transition times when territories are being set up. But such occasions are short-lived and almost always decided by bluff and threat, with the territorial owner at a clear advantage. Territoriality includes a tendency both to establish one's own piece of real estate and to respect the ownership of others. In fact, territorial ownership often confers near invincibility upon the proprietor. The boundaries may be remembered by experience, but respect for them is biologically rigid.

If we were truly territorial, fighting over space would be greatly reduced, especially once the boundaries were established. It is precisely because we lack much biological respect for territory that we fight so often over it. We want our own space, sometimes enough to fight for it, but at the same time, we are disinclined to respect very profoundly the territory of others. In most of the Western world we commonly stake out spatial claims, using picket fence, stone wall, barbed wire, printed sign, or locked door, all of which are themselves testimony not so much to territoriality but to its absence. And in doing so, we are following our culture, not our genes. Because these artificially inspired boundaries lack the support of biological evolution, they are rather insecure and subject to constant dispute. When cultural evolution presents situations for which biological evolution has not prepared us, look for trouble.

Despite the likelihood that human beings are not biologically territorial, let us reverse our field for a moment and assume for the sake of argument that we *are* territorial animals after all. A territorial biology would likely produce some sort of regular spacing between individuals and almost certainly, between families. In some cases, this could be accomplished by the widely spaced distribution of farms in the Midwest or the oppressive regularity of a typical suburban subdivision. Both patterns occur today, and the distances involved vary greatly. But, if there is an appropriate pattern of human spacing that best satisfies our genetic needs, it may be fifty feet apart, as in modern Levittown, or fifty miles, as in parts of Alaska. Not knowing, we cannot arrange our living to coincide with such possible needs. Moreover, the existing diversity of patterns suggests that even if we had such needs, they must be violated by the vast majority of living arrangements, since the variation is so great and the determining factors seem unlikely to be biological appropriateness.

The human genetic makeup—whatever it may be—is poured, like concrete, into molds of territorial use that are almost entirely determined by political, social, economic, and possibly random factors. But unlike concrete, genetic influences are not infinitely malleable. They cannot comfortably assume an infinite array of configurations, and if forced into inappropriate shapes, they may be stressed to the point of cracking. If we have any genetic propensity for territorial organization, then it is likely to be distorted and stressed by the wide diversity of contemporary human culture. Alternatively, if we lack such a propensity, then as we have seen, trouble also follows because we lack the stability that comes with biologically mandated adherence.

When it comes to questions of territory, the real crunch of culture upon biology could be expected to occur, once again, in our large urban areas. Just as a genetic tendency for social ranking, even if present, would be difficult to maintain in a big city, any purported territorial tendencies must be greatly strained in the urban crush. The motives for city dwelling are varied and complex. But economic factors clearly provide a major inducement to the urban dweller, with subsidiary incentives of excitement, novelty, sociocultural diversity, and simple inertia (having been born into the situation). It seems likely that even if territorial needs exist within us, powerful cultural factors with immediate appeal could readily seduce us to act contrary to them.

If human aggression derives in part from our culture acting upon a relative *absence* of biological characteristics, what sort of environmental determinants of aggression can be postulated? In *The Republic,* Plato sought to identify the proper form of human government, based on the underlying nature of the human soul. Although successful as philosophy, such efforts have not been notably successful in practice, perhaps because when it comes to human preferences for social organization, there is no single, underlying human soul. Our natures may be as insubstantial as the famous shadows dancing on the wall of Plato's cave.

If indeed we lack a genetic tendency for establishing any one particular kind of social system, our various forms of social organization must be strongly influenced by culture, perhaps entirely so. Human political systems are very diverse, ranging through tribalism, absolute monarchy, military dictatorship, republican democracy, fascism, socialism, and communism. Family structure, in turn, runs the full gamut from the private nuclear family to extended clans, and social patterning runs from densely urban and sedentary to pastoral and nomadic. Perhaps most important of all is the basic, diffuse social fabric of personal relationships, from the superstructure of political organization to the foundations of daily, family life. Here, the diversity of human relationships and organizations is truly staggering.

This varied social architecture has almost certainly developed without corresponding blueprints in our genes. For systems as diverse as communism and democracy, the guiding force has been a specific social, economic, and political philosophy, held initially by a few individuals. Such ideas have either prospered or withered, depending on the cogency with which they were expounded and the social, economic, and psychological climate among each group of people that was exposed to them. Social infrastructure, by contrast, reaching down ultimately to basic family relationships, is probably more resistant to alteration by a momentarily attractive ideology. In China, the human family has proved resistant to early Maoist efforts at drastic modification, just as in the United States human compassion has resisted the New Right's efforts to repeal New Deal entitlements and organize personal relations around selfish disregard of others. Although the fundamental social fabric of human behavior experienced by nearly all people is likely to be deeply biologic, the superficial patterns encountered by most of us seem to be largely a result of inertia and conformism; that is, we practice the social organization that our parents and relatives did, and that we see exhibited by our contemporaries, perhaps with a dollop of biological assistance.

To some extent the mosaic of social organizations exhibited by human beings is probably adaptive as well. We might therefore expect that the optimum social organization differs among tropical rain forest, prairie, seashore, and arctic dwellers, although the ideal arrangement for each particular case cannot be identified. But regardless of the forces producing specific patterns of social organization, such patterns once established would probably be continued by the joint actions of simple inertia and conformism (provided, of course, these patterns are not so grossly maladaptive as to produce extinction of the group, or so uncomfortable as to generate revolt).

Regardless of the methods of transmission, and irrespective of their adaptive values, it is likely that a large proportion of human social institutions lack much support from our genetic makeup. We can therefore anticipate a fair degree of instability in such systems, and once again, psychologists tell us that such instability is a prime breeding ground for aggression. In short, our social organizations (culturally evolved) seem likely to break down frequently since they lack corresponding support from our genes (biologically evolved). The result? More aggression.

It seems likely that we lack a genetic propensity for any particular form of social organization, and that in fact we have biologically evolved a capacity for utilizing many such cultural systems, just as we speak many different languages. Culture itself is a kind of specialization; we have specialized in being generalists. By not restricting us to a limited number of systems, biological evolution has provided one of the essential keys for successful exploitation of a great variety of habitats. Nonetheless, insofar as we are jacks of all trades but masters of none, we would still continue to lack the stability born of strong biological support for a particular way of life. Honeybee society, for example, is well organized and peaceful: with a place for everyone and everyone occupying his or her place. The workers work, the queen lays eggs, and the drones drone on. The fact remains that social breakdown is a prime ingredient in aggression, and any human culture—developed for whatever reason—that lacks a firm genetic substrate such as the honeybee enjoys can be relatively susceptible to such breakdowns.

All this should not be mistaken for an argument in favor of any particular type of social, political, or economic organization, in the forlorn hope that such an organization would be more congenial to our genes. In fact, it is exactly the opposite: since our genes apparently do not specify a social system for Homo sapiens, we are free to choose what we wish, or more likely, to suffer whatever is imposed on us. As the Grand Inquisitor pointed out in *The Brothers Karamazov,* freedom of choice can be an awful burden, and when it comes to social systems, the weight of that burden is especially real and can be especially troublesome, since there may be no system in which we are truly comfortable. Like the expatriate or the dedicated internationalist who can live with equal ease in a variety of countries but never feel that he or she is at home anywhere, Homo sapiens can live in a great variety of social systems without ever really being at home.

As we did when considering human territoriality, let us now

make the opposite assumption; namely, that we *do* possess a biologically evolved predisposition for some particular pattern of social organization. Here again, as with territorial behavior, the enormous heterogeneity of human cultural arrangements guarantees that many, if not most of them, are somehow inappropriate for us. It is significant that whereas the disruption of social organization may well be a major cause of aggression, much human violence seems to be directed toward disrupting existing cultural subsystems, rather than to be a result of their disorganization. Both of the above suppositions could act independently, and both lead to the same result. If we lacked genetic tendencies in support of our various cultural systems, we would clearly also lack any great reluctance to destroy them. On the other hand, if we possessed a genetic predisposition for some particular type of social organization and were denied fulfillment of it because of economic or ecologic conditions, or just simple historical accident, we might be inclined to rebel. In either case, aggression would ultimately be very likely: damned if we do and damned if we don't.

For a final example of this paradox, consider some consequences of the fact that we evolved in the tropics. The fossil record and our basic physiology both strongly suggest this. Our cold tolerance, for example, is extremely poor. Unclothed and without some basic technology, we could not survive outside the tropics. Yet clearly we do. In fact, we have spread over the globe, experiencing the widest geographic range of any species, and we have achieved this fantastic expansion by virtue of our master adaptation, culture. Human culture has thus allowed us to explore otherwise inaccessible places, settle in otherwise inhospitable lands, and fill a variety of ecological niches ranging from completely vegetarian to meat- and even insect-eating. Culture has been our necessary support and without it we would be lost; indeed, we would cease to be human. It frequently happens, however, that in relying so heavily on something that it becomes essential, we surrender some autonomy to it. We profit greatly from language, for example, and rely on it; at the same time, as linguist Benjamin Whorf has emphasized, our perception of the world is colored by the linguistic rules and vocabulary that we experience. We take advantage of mechanized agriculture to feed our immense population, and in the process, become dependent on pesticides, and millions die when the rains fail to come on schedule.

Assuming now that we possess some biological basis for our behavioral systems, we might also assume that these systems would be most appropriate for use in the tropics, where most of our early biological evolution took place. Our innate tendencies, under the watchful eye of natural selection, would then be attuned to systems that are adaptive for a tropical existence, whatever systems they may be. But as our culture began its rapid evolution, we developed the capacity to exist in regions far removed from our ancestral habitat. We rode our culture to mountain and desert, to subtropical and ultimately temperate and arctic regions. In so doing, we were probably forced by the exigencies of new and sometimes rather strenuous environments to employ cultural practices that were at least minimally adaptive. We had no choice but to deny, or at least ignore, any tropical biology that might still whisper within us. So the real culprit may be not culture, or biology, but the human wanderlust that forced us to adopt cultural styles that may be at variance with our biology.

Arnold Toynbee has suggested that the technological development of temperate zone people is due to the stimulating effect of seasonal change, combined with a certain degree of environmentally imposed necessity. We might expect to find Homo sapiens increasingly dependent on culture as we inhabit environments that are increasingly removed from our biological heritage. In this regard it would be interesting to see if temperate and arctic societies exhibit higher levels of neurosis, personal alienation, and aggression than do their counterparts in the tropics.

FIGHTING and winning makes an animal more likely to fight again. The same has been proposed for human beings. If so, it might also be that the vicarious experiencing of aggression—in other words, seeing or hearing it—would reduce inhibitions and foster aggressiveness in the spectator, just the opposite of the predictions generated from a strictly ethological perspective. Throughout history, human cultures have shown great ingenuity and expended considerable effort providing spectacles of aggression for their members. The ultimate motivation behind these performances may be the belief that vicarious experience of aggression would have a cathartic effect, like the Aristotelian concept of Greek tragedy, causing the observer somehow to discharge pent-up energies and "get the anger out." Perhaps it was expected that such experiences would render him or her less likely to threaten society with violence. Hence, the Roman fondness for "bread and circuses."

On the other hand, public display of aggression may also satisfy a profound need in our species, and not just in our rulers or in those poised to profit financially from such events. Primitive societies often hold ceremonial gatherings at which aggressive posturing takes place, with dancing, fighting, and even killing, often of sacrificial animals. In the worst case, a vicious circle ensues, with aggression feeding on itself as it demands public representation, which in turn generates more personal aggressiveness, which requires yet more public representation, and so on. Western society has produced the Roman gladiators, Christians eaten by lions or nailed to crosses, bullfights, public executions of all sorts, boxing matches, demolition derbies, and football games. With the advent of warfare, most recently abetted by radio and television, the details of organized human aggressiveness have been made apparent even to the noncombatant. Unlike the above examples, however, televised warfare is presumably not designed specifically for home entertainment. And in fact, televising of the Vietnam War seems to have increased nationwide repugnance for that conflict, whereas by contrast one of the dangers of artificial televised violence may be that it insulates the viewer from the real consequences of violent aggression. With that war finally ended, and with the result unsatisfying to boot, the American public was then free to indulge its fantasies of martial triumph, vicariously, through Rambo and other safe, celluloid escapades of wish fulfillment.

Psychologists have also proposed "frustration" as a major environmental cause of aggression. Webster's Dictionary defines frustration as "being prevented from achieving an objective or from gratifying certain impulses and desires, either conscious or unconscious." If we postulate a genetic tendency for certain behavior in human beings, such as territoriality or the establishment of a particular kind of social hierarchy, culture emerges as a major frustration in that it must often keep us from achieving our unconscious objectives. If Freud and Hobbes are correct, and civilization acts to frustrate and impede our otherwise unacceptable tendencies—indeed, if civilization is only possible because of this impedance—then culture may well be in the ironic position of causing aggression by the very act of preventing it. Again it seems that we just can't win.

One of the great benefits of culture, and indeed, one of the primary reasons for its existence, is that it empowers human beings to fulfill many of their desires. A certain degree of mastery over nature, or at least, temporary insulation from it, is made possible by technology which provides food, clothing, and shelter. Some leisure time is also generally assured, as well as numerous opportunities for self-expression through artistic media and (for the fortunate) through one's work. Theoretically at least, anyone with sufficient inclination and capacity to inquire into the workings of the natural world can become a scientist. Freed, in part, from those biological constraints that tended to make life nasty, brutish, and short, human beings have become the most playful animal on earth: Dutch historian Johan Huizinga once proposed the moniker *Homo ludens,* Man, the player. All this, and more, we owe to culture. It has been proposed that the relatively nonviolent African bushmen owe their low level of interpersonal aggression to a penchant for vigorous, energetic playfulness. And among animals, a playful mood is a distinctly nonaggressive one.

Our desires for social companions, of a particular sort, at particular times, doing particular things, can to some degree be satisfied by our social systems. Human culture may even be defined as a complex nongenetic system that satisfies human objectives, impulses, and desires. So despite the restrictions that it imposes, culture is also, ironically, an enormous nonfrustration device.

But beyond the satisfaction of such basic animal needs as food, warmth, sleep, elimination, reproduction, and sex, can anyone identify what a human being really wants? Culture doubtless provides much that we could not otherwise achieve. And yet it is no contradiction to point out that culture also generates many desires that must remain unfulfilled; that is, it also promotes frustration of a different sort. While satisfying many needs, it creates want. Someone living in today's highly evolved technological culture is likely to be more full of "wants" than his or her primitive ancestor, living at less remove from biology. Modern Homo sapiens, living atop an enormous superstructure of cultural opportunities and expectations, often finds him and herself assailed by vague stirrings of dissatisfaction. A potent source of such feelings is that old bugaboo: envy.

Modern culture has no monopoly on producing such feelings. It is likely that one of our early human ancestors, eyeing the results of a successful hunt, dearly wished that he and not his colleague was chewing on the giraffe's bloody haunches, or that she could have as easy a labor and delivery as her neighbor. And we have probably coveted our neighbor's wife (or husband) since the pair bond was first liberated from the genetic inflexibility found in most animals. But the scale at which modern culture now generates such emotions is probably unprecedented in the evolutionary history of our species.

Advertisement and the mass media also tease us with the images of some of our species's most attractive products (both human and material). Little wonder frustration is rampant. Once human culture progressed beyond the point of mere sustenance, we began surrounding ourselves with things. The keeping of possessions can easily lead not only to aggressive competition, but also to aggression via frustration. Conspicuous consumption is more than just a cliche for American status seekers. It is usually assumed that people keep up with the Joneses in order to raise their status; by contrast, our concern is not so much with the motivations of the conspicuous consumer as with his or her impact on that neighbor. One of the most subtle and powerful ways in which culture generates frustration is by exposing individuals to glimpses of another's possessions and accomplishments. By generating desires that cannot readily be gratified, and that may never be, culture is frustrating no less than gratifying.

Culturally induced frustration is especially apparent in most of the world's big cities, where rich, poor, and middle class almost literally rub shoulders—and sensibilities—every day. It is probably even more potent among many Third World nations, where the "revolution of rising expectations" is fueled by newfound awareness of the affluence of other people. Poverty is never easy to bear, but it was more tolerable before magazines, radio, television, and tourism made the impoverished aware of how badly off they really are: not just in absolute terms, but compared to the relative affluence of the Western world.

FINALLY, let us consider war. The relationship between war and aggression remains moot: war may be a *result* of aggression, simply its best organized and most deadly manifestation. Or conversely, war may be a *cause* of human aggression, the product of high-level governmental decisions which are then translated into personal resolve and emotion by the various stimulative and exploitative devices at society's command. Viewed either way, culture makes a major contribution to aggression by providing numerous opportunities for conflict between societies.

Alone among all animals, we provide the unique spectacle of individuals killing other members of the same species in order to "convert" them to some cultural practice. The cultural distinctions between people, whether adherence to different religions, economic systems, or ways of gaining a living (e.g., farming vs. herding) have always loomed large in humankind's reasons for going to war. We are all one species underneath, but we come packaged in many different cultures, thereby providing numerous options for distinguishing Us from Them. National identification, which typically aspires to be family and kinship identification writ large, facilitates both a defensive sensitivity and a dehumanization of the opponent.

Despite relatively high levels of aggression, animals only rarely kill or even seriously injure others of the same species. Recent field studies have shown that free-living animals are not nearly as idyllic as had been believed until the mid-1970s: wolves occasionally kill other wolves, lions sometimes kill other lions—typically, outsiders—and infanticide is quite frequent, especially when a male takes over an existing harem. In general, however, no living species practices the degree of intraspecies killing that we do, as measured either by its frequency or its ferocity. Ethologists have long emphasized that aggressive restraint among animals is typically achieved by a combination of behavior traits, nearly all of which have been evolved through natural selection and which accordingly rely on genetically coded mechanisms. When two rivals meet in an aggressive situation, bluff and threat generally ensue. In fact, the vast majority of such contests are immediately settled in this manner, with no recourse even to physical contact, let alone injury. By displaying themselves to best advantage, each contestant tries to impress the other with his prowess, so that the intimidated rival acknowledges inferiority or simply goes away.

Each species generally possesses a unique set of behaviors used for such occasions. The common features usually include exposure of teeth, claws, or whatever weapons the animal may possess, plus making itself appear as large as possible. Among fish, this may involve puffing up, spreading the fins, and orienting laterally to the opponent, while mammals commonly make their hair stand on end; all create the illusion of greater size. It is clearly better for disputes to be settled this way than by a prolonged, exhausting fight, with the threat of possible injury or death.

Should the two rivals appear evenly matched, a ritualized form of fighting may ensue, often resembling stylized posturing more than an actual struggle. Fish of many species will lock jaws, pushing and pulling each other in a combined effort to assess the others' strength while emphasizing their own. This is the basis for the famous and titanic struggles of bull elk or deer of many species, in which the males lock antlers and strain mightily against each other.

If the intent was to kill, such antlers could be used much more effectively against the unprotected sides of a rival. But this almost never happens. Instead, the competitors avoid such potentially lethal attacks, turning aside from their opponents' vulnerable flank and waiting for the opportunity to lock horns in ceremonial struggle once again. So ritualized are these behaviors that the pioneer European ethologists who have studied them in detail describe them as "tournaments" rather than true fights. Marine iguanas push against each other, locked forehead to forehead until the loser acknowledges defeat by dropping to his belly while the victor remains stiff-legged and threatening until his defeated rival scurries away.

Significantly, when males have lethal weapons and females do not, the latter typically lack the stylized aggressive tournaments so characteristic of the former. And those species that lack lethal weaponry altogether also tend to lack inhibitions against attacking one who is vulnerable.

Rattlesnakes fight frequently, and they are not immune to rattlesnake venom. (They are able to eat their own venom-injected prey only because the poison is broken down by the digestion; if bitten by another rattlesnake, they can die.) It is thus interesting that fighting rattlesnakes scrupulously avoid biting each other.

Instead, they engage in peculiar, highly stylized wrestling matches in which each one tries to push the rival over on its back. The two adversaries face each other with their heads and about one third of their bodies raised vertically above the ground, bobbing and weaving, while sometimes rubbing their ventral scales together. In making contact and straining against one another, they may raise themselves several feet above the ground, possibly giving rise to the caduceus symbol of the medical profession in the process. The victor pins the vanquished with the weight of his body for just a few seconds, and then the loser glides away, defeated but unbitten and very much alive.

Other animals with similarly lethal weapons can be counted on to refrain, typically, from using them against members of the same species. Thus, the Arabian oryx—equipped with long razor sharp horns—uses these weapons to push sideways and harmlessly against a rival. And giraffes, with strong and potentially dangerous hooves, harmlessly butt heads during aggressive confrontations. But just as the rattlesnake does not hesitate to use its venomous fangs against a wood rat, the oryx will use its horns and the giraffe, its hooves, against lions.

It is to an animal's selective advantage if it can efficiently dispatch either its would-be predator or its prey. But there is no real advantage to killing members of one's own species if the same result can be accomplished more easily and with less danger. After all, in many cases the loser in a "tournament" situation may well be a relative. If so, killing such a rival would be a Pyrrhic victory indeed, likely to be discouraged by kin selection. If success can be achieved without death or even severe injury to the opponent, then the victor could even gain some benefits from the other's company, without having to pay the price of fitness-lowering competition. Finally, interactions with prey or predator are comparatively one-sided as opposed to the double-edged sword of murderous intraspecies fighting. Selection for lethal attack on another species member could easily boomerang, creating the chances of either contestant being killed if both possess the appropriate genes, analogous to mutual defection in the Prisoner's Dilemma. Students of mathematical game theory have also demonstrated that under certain conditions, behaving as a "dove" can be more successful than behaving as a "hawk," even if hawks prevail in dove-hawk contests. (This is because hawks, as they become more numerous, are more likely to encounter other hawks and then to suffer damaging fights, whereas doves are more likely to live to fight another day.)

The avoidance of within-species killing, although not universal, is relatively well established for most animals, and is usually achieved by elaborate biologically evolved behaviors that are especially pronounced among those animals with a greater potential for lethality. A rabbit or a robin is clearly less likely to assassinate a colleague accidentally, or to be killed in turn, than is a leopard, eagle, or rattlesnake. Lethally armed animals, not surprisingly, have evolved effective and relatively harmless behavioral substitutes for murderous aggression; namely, displays and tournaments. Even when such measures are seemingly unable to prevent a "real" fight, a last fail-safe mechanism is usually available. For example, Konrad Lorenz has pointed out that the most "vicious" fight between two wolves in nature generally will not result in the loser being killed. Once a victor has clearly emerged, the loser does something that automatically inhibits the winner from pressing home its attack. Among wolves this may involve turning the neck to expose the soft underside, exactly the most vulnerable part of the animal's body; a similar response is seen when domestic dogs roll over on their backs, exposing their belly to a larger, dominant dog, or to a human being. Instead of killing his hapless opponent, the victor may then snap his jaws harmlessly in the air. Once another animal shows itself to be subordinate through this stereotyped behavior, it is unlikely to be killed. In short, the capacity for killing has generated controls which prevent its misuse. (Actually, we now know that wolves do occasionally kill each other, mainly when individuals from different packs come into conflict; the basic principle of restraint still holds, however, at least for fights within the wolf pack.)

In most cases in which animals have been found to kill other members of their species, the individuals had been in captivity. Often they are either prevented by the poverty of their artificial environments from exercising their full behavioral repertoire or they are confined in such small cages or tanks that simple retreat, the most common recourse of the defeated animal, is denied them. Rattlesnakes do not kill other rattlesnakes. But throughout recorded history, men have killed other men, and often, women and children as well. Perhaps the most persistent sound, echoing through our own evolutionary past as well as recent times, has been the beat of war drums.

The question of morals really isn't at issue here, especially since morality and ethics are basically society's effort to substitute cultural control of behavior for the biological control that we generally lack. The rattlesnake is responding to its genetic rather than its ethical system when it refrains from killing a rival. Despite the moral authority of the Mosaic prohibition against killing other human beings, we conveniently ignore that commandment without denying our biological heritage. Society finds it useful to inhibit killing when it might disrupt the smooth functioning of human social organization. But when society considers killing to be in its best interests, generally the killing of deviant individuals (executions) or of perfectly normal members of another society (warfare), the culturally invoked restrictions are easily lifted. Just as there has been virtually no effective biological evolution among human beings during the past two thousand years, there has been no effective moral evolution either. By A.D. 600, our species had already displayed the wisdom of not only Moses and Christ, but also Lao-tse, Confucius, and the Buddha. In the twentieth century, we had Hitler and Stalin . . . so it is not at all clear whether progress has been made. On the other hand, during the same interval we had enormous "progress" in weaponry, from the lance and broadsword to nuclear missiles.

One fundamental question remains to be answered: why do we lack effective, guaranteed, biologically evolved inhibitions against killing? The answer lies once more in the disparity between our biological and cultural evolution. In fact, it provides one of the clearest examples.

We lack genetically mediated killing inhibitions because natural selection didn't have much reason to endow us with any. After all, a naked, unarmed human being really isn't a very dangerous adversary to another, similar human being, unless he or she has had special (modern) combat training. It is extremely difficult for one person to kill another using only the devices with which biological evolution has equipped us. Our hands and feet are very inefficient weapons and our reduced teeth in their muzzleless, flat face present little threat of a killing bite such as found in the dog, cat, or weasel families, or even the modern reptiles and fish. Thus, throughout most of human evolution we were like the robin, rabbit, or female deer: lacking weapons capable of killing our fellows, we also lacked inhibitions against such killing, since we were virtually incapable of it anyhow.

Then came the rapid explosion of cultural evolution and the successive discovery of stone, club, knife, spear, blowgun, bow and arrow. These weapons probably first served to facilitate killing of prey or defense against predators. But defense and attack against other people were probably not far behind. Intraspecies fighting became the main function of human armaments, and with a great rush we invented swords, muskets, cannons, machine guns, poison gas, battleships, tanks, bombers, submarines, guided missiles, and finally, nuclear weapons. When the stag evolved his antlers or the rattlesnake its poison fangs, time was abundant—measured in millions of years—and the biological world stayed relatively unchanged for millennia. Each stage of development was accompanied by the corresponding evolution of genetically mediated behaviors carefully titrated to the biologically evolved equipment at hand. Otherwise, elk and rattlesnakes would probably have gone extinct long ago. The extraordinary development of human weaponry has occurred in a span of *thousands* of years, and our really crowning achievement in fiendishly destructive capacity is a product of the mid-twentieth century alone.

Albert Einstein once noted that whereas he had been taught that modern times began with the fall of the Roman Empire, in fact they began with the fall of the bomb that destroyed Hiroshima. In any event, such a time scale is much too short for natural selection to have operated effectively, or indeed, at all. We possess armament that is far more deadly than the most dangerous animal, but since it has developed by cultural rather than biological evolution, we lack the necessary behavioral controls that natural selection would have assured. Biological evolution would never have allowed such a souped-up hot rod to leave the assembly line without a good set of brakes. Biologically we are still rather innocuous apes, while our unrestrained culture makes us the greatest potential (and actual) killers the world has ever known.

It is conceivable that human beings are somewhat susceptible to appeasement or subordination behaviors by their fellow humans. And this may even have a weak genetic basis. We are generally less likely to kill a submissive, defeated individual who kneels at our feet. But the bowed head of the prospective victim has never in itself stayed the guillotine or the headsman's axe; human history abounds with the overt killing of supplicants. Admittedly, we do show a certain reluctance to kill women and children, somewhat preferring to slaughter our fellow *man*. This is probably of modest selective value, since as we have seen, males of most species compete with each other for access to females, as vehicles for perpetuating their (the males') genes. Killing the men and raping the women is thus a partial expression of our biology. In fact, many cultures take advantage of this apparent inhibition against killing women and children by using them as emissaries of peace or surrender, and also by distributing them as booty. But witness the slaughter of entire Cheyenne families by the American cavalry at Sand Creek, or the equivalent butchery at My Lai.

Maybe if we only possessed the right repertoire of subordination behaviors and automatic responses to them, then all would be well. But unfortunately even the unlikely development of biological controls to match our galloping culture would avail us little, because technology has provided us with increasingly efficient weapons that do their deadly work from increasingly greater distances. Subordination behaviors are only effective at the personal level: even if a villager is engaging in an exceptionally effective appeasement display, she cannot be noticed by the bombardier 20,000 feet up, or by the politician on another continent with his finger on the nuclear trigger. The whole trend of developments in human weaponry has been for effectiveness at ever increasing distances, from the club at three feet away to the cannon with a range of several miles, to the ICBM at 10,000 miles. This in itself makes biological inhibition a virtual impossibility.

Is there a solution? It is often easier to identify a problem than to solve it, and this seems particularly true of human aggression. Our culture, by the breathtaking speed of its advance, has placed us in a very dangerous position. But the capacity for culture is part of our genetic heritage and we could no sooner give it all up than forego use of our thumbs because they occasionally get hit by a hammer. Just as we must learn instead to use a hammer wisely, we had better develop proper control over our cultural productions. Perhaps we should also accept that certain things are too dangerous for us to use safely, or even to keep around at all. By their rambunctious primitiveness, human beings sometimes seem like incautious little children, who must be protected from their own immature inclinations; but they are the adults now, and there is no safe medicine chest in which we can safely store the household poison.

It should be emphasized that evolutionary considerations do not lead irrevocably to an accusatory posture in which Homo sapiens is castigated for being irrevocably tainted with a kind of biologically based original sin. First of all, nothing in human behavior is irrevocable. Although we irrevocably carry the imprint of evolution, this does not mean that we must irrevocably behave in any particular way. And second, our biological failings are more those of omission than of commission. We are threatened more by the genetic traits that we lack than by those we possess. Species self-loathing has been a surprisingly popular passion among human beings, and it has led more to an occasionally gratifying intellectual hair shirt than to useful insights. Among such protestations of disgust, some of the most widely disseminated have accompanied the recognition of our likely carnivorous ancestry, as reflected, for example, in these outraged observations by anthropologist Raymond Dart, in his fascinating and controversial *Adventures with the Missing Link:*

> The creatures that have been slain and the atrocities that have been committed . . . from the altars of antiquity to the abattoirs of every modern city proclaim the persistently bloodstained progress of man. He has either decimated and eradicated the world's animals or led them as domesticated pets to his slaughterhouses.

Mistaking predatory behavior for intraspecies aggression and war, Dart discerned the origin of human warfare in those torrents of animal blood:

> The loathsome cruelty of mankind to man . . . is explicable only in terms of man's carnivorous and cannibalistic origin. The blood-spattered, slaughter archives of human history from the earliest Egyptian and Sumerian records down to the most recent unspeakable atrocities of World Wars I and II, accord with universal cannibalism . . . in proclaiming this common blood-lust differentiator, this mark of Cain, that separates man dietetically from his anthropoid relatives.

Aldous Huxley expressed a similar, and not entirely unjustified feeling:

> The leech's kiss, the squid's embrace,
> The prurient ape's defiling touch:

And do you like the human race?
No, not much.

Let's be fair to our species; after all, it is the only one we shall ever be. We are neither gods nor devils, and with a bit of insight, we can even aspire to that state of secular grace urged upon us by Albert Camus, at which we shall be neither victims nor executioners. There may well be too much feisty competitiveness built into the human spirit for us ever to lie down together like lambs. At the same time, there may well be just enough evolutionary leavening in the form of enlightened self-interest, for us to reach a middle ground in the human tempest, something between the brutish barbarism of a Caliban and the lofty cerebralism of a Prospero. The word "enemy" derives from the Latin *in* (not) plus *amicus* (friendly), and it implies a state of hostility, often involving hatred for one's opponent. By contrast, the word "rival" originated with the Latin *rivus* (stream), and literally means "one who uses a stream in common with another." Rivals are therefore competitors; this much may be inevitable. They need not, however, be enemies.

Near Southbridge, Massachussetts, on the Connecticut border south of Worcester, there is a lovely lake with the extraordinary Mohican name of Chaubunagungamaug. In English it means "You Fish on Your Side, I Fish on My Side, Nobody Fishes in the Middle: No Trouble." The early residents of Lake Chaubunagungamaug, we may assume, were rivals but not enemies.

It is not too late for human beings to treat each other as rivals rather than enemies. It is too late, however, for biology to achieve this transformation, acting alone. The time is too short, and the need too urgent. We have traveled too far down the road of culture; having surrendered ourselves to its power, excitement, and uncertainty, we must ultimately find our salvation in that route as well. Once we disrupt a natural system—for example, by planting a field of corn—we are forced to keep relying on human input through cultivation, weeding, irrigation, and possibly even pesticides and herbicides, just to keep that system from falling apart. Similarly, we desperately require a stringent cultural system that will respect the exigencies of our evolutionary past while protecting us in our perilous, but irrevocable, departure from our biology in the present and future.

CHAPTER EIGHT

THE NEANDERTHAL MENTALITY

CAVEMAN CONSCIOUSNESS IN THE NUCLEAR AGE

When you look into the abyss, the abyss also looks into you.
—FRIEDRICH NIETZSCHE

Looking into the nuclear abyss, we can see nothing less than the end of the world, something that prophets have warned about for millennia, but which now—thanks to the possibility of worldwide environmental catastrophe known as nuclear winter—has at least become a potential reality. And looking into ourselves, we find a twentieth-century caveperson, wielding unheard of powers but still a primitive at heart.

"The splitting of the atom has changed everything but our way of thinking," wrote Einstein, adding, "and hence, we drift toward unparalleled catastrophe." Now, almost forty years later, that "drift" has become a powerful tide. We are confronted with something much harder than a mere physics problem: a *psychology* problem wrapped up in modern politics, driven at least in part by our evolutionary past. As Einstein himself recognized, psychology and its public manifestation, politics, are much more difficult—and important—than physics. It is crucial, therefore, that we begin to understand our way of thinking (and sometimes, not thinking) about nuclear weapons. We might also inquire into why our thought processes haven't changed, even though with the splitting of the atom, everything else has.

Nothing concentrates the mind, observed Samuel Johnson, like the prospect of being hanged in the morning. Similarly, nothing ought to concentrate the human mind like the growing prospect of nuclear war. But unfortunately, today's caveman is as ill-equipped to deal creatively with modern weapons as he or she is well equipped and often eager to swing a club—be it zebra bone or nuclear armed.

There is nothing unusual about possessing lethal weapons; the porcupine, for example, has prospered with its dangerous quills. But whereas a porcupine's mental and physical traits developed synchronously, ours are increasingly out of phase. When it comes to defending itself against predators, a porcupine relies on its sharp quills, and the combination is effective, because neither its behavior nor its quills have changed in millennia. The porcupine's quills are appropriate to its reliance on them because the two evolved in concert, the formidable weapons along with the appropriate instinctive and readily learned porcupine behavior patterns.

Not so, however, for modern human beings. It seems likely that for nearly all of our evolutionary history, we were rather porcupinelike ourselves; that is, our behavior and our capabilities were more or less in tune with one another. But in the last few thousand years, this cozy relationship has gone awry and nowhere is this disparity more spectacular or more dangerous than in the realm of nuclear weapons. Manipulating the brave new world of culture—in particular, the brave new nuclear weapons—is the same old biological Us. A Neanderthal has his finger on the button.

But, we reassure ourselves, no rational person would start a nuclear war, and can't we count on the rationality of Homo sapiens? On the contrary, "Only part of us is sane," wrote Rebecca West,

> only part of us loves pleasure and the longer day of happiness, wants to live to our nineties and die in peace, in a house

> that we built, that shall shelter those who come after us. The other half of us is nearly mad. It prefers the disagreeable to the agreeable, loves pain and its darker night despair, and wants to die in a catastrophe that will set life back to its beginnings and leave nothing of our house save its blackened foundations. Our bright natures fight in us with this yeasty darkness, and neither part is commonly quite victorious, for we are divided against ourselves . . .

As the gap between the hare and the tortoise has grown, the gap between human survival and oblivion has narrowed. More than twenty years ago, social psychologist Charles E. Osgood coined the phrase "Neanderthal mentality" in his influential discussions of the psychology of the arms race. Although clearly it is not anthropologically accurate (Neanderthalers may not even have been on the path of human descent), the term speaks eloquently of tendencies that are widespread and primitive. It is also appropriately pejorative. Unfortunately, the Neanderthal mentality is all too evident when erstwhile Homo sapiens confronts, or fails to confront, the problems posed by nuclear weapons.

What, then, is the Neanderthal mentality when applied to nuclear weapons? Basically, it is the tendency to employ prenuclear mental processes to a totally new, nuclearized world. To evolution, 1945 is less than yesterday. Despite a number of temporary setbacks, and a growing array of problems, the Neanderthal mentality served us well, for 99.999 percent of our evolutionary history, during which time it generated behavior that led on balance to social and biological success. But thanks to Einstein himself, and the Manhattan Project, the scene has suddenly changed. Thanks to biological evolution, however, the actors keep reading from the same outmoded script.

There is a problem, then, not only with the nuclear "hardware" but also with the human software. As Harvard negotiator Roger Fisher has pointed out, the mere possession of nuclear weapons does not in itself guarantee unparalleled catastrophe. France and Britain, for example, are both nuclear armed and have been mortal enemies for centuries. Yet strategic planners in London do not lie awake nights, anxiously anticipating a preemptive first strike from Paris. Closer to home, the United States Army, Navy, and Air Force are all immensely powerful, antagonistic, and armed to the teeth with nuclear weapons, yet they restrict their competition to the football field and Congressional appropriations committees.

Here are four aspects of the Neanderthal mentality that encapsulate the nuclear dilemma as resulting from the conflict between culture and biology.*

The first concern revolves around the question of aggressiveness and getting one's way in a pushy world. There is primitive appeal, for example, in the notion that we make ourselves safer by placing our opponents at risk. For generations, this may well have worked. If so, it was likely promoted by natural selection. But it is no longer appropriate to a world in which security, if it is to exist at all, must be mutual. As Roger Fisher points out, there is an American at one end of a rowboat and a Russian at the other, and the strategy of each seems to be that he will make his end of the boat more secure by making the other fellow's end more tippy. Following this "logic," we each deploy missiles theoretically capable of destroying the other side's missiles; this, in turn, makes Them less secure and is therefore supposed to make Us more secure. But in fact, the more nervous and jumpy we make them (and/or they make us), the more likely it becomes that one side or the other, fearing that it will not be able to retaliate if attacked, winds up attacking instead. And the more each side threatens the other, the more likely it becomes that either side will mistake a false alarm for the real thing. As a result, the frail craft we call the earth is more and more likely to capsize. And yet, to the Neanderthal, the ritual of reciprocal risk makes a primitive kind of sense.

For a very long time, having *more* made our primitive ances-

*For a much fuller development of the Neanderthal mentality applied to nuclear weapons, see *The Caveman and the Bomb: Human Nature, Evolution and Nuclear War*, by David P. Barash and Judith Eve Lipton, McGraw-Hill, 1985.

tors safer. As nations feel increasingly insecure in the nuclear age, then, the answer seems simple enough: more clubs, more warriors, more bows and arrows, more guns, more tanks, more bombers. But the circumstances have changed, and getting more does not make anyone safer; in fact, quite the opposite. But try telling that to today's nuclear Neanderthal. The more insecure he becomes, the more tightly he clutches the very sources of that insecurity, getting more and more weapons, and in turn getting more and more insecure when the other side follows suit, all the while looking more and more like a muscle-bound idiot in a Chinese finger puzzle.

Closely related to "more is better" is "fewer is worse": fear of having fewer nuclear weapons than one's opponent, no matter how meaningless the so-called nuclear balance. Accordingly, despite the fact that the United States has always been ahead in the nuclear arms race (and still is ahead today) we have continually been stampeded by an array of fictitious gaps; since the early 1950s, we have been subjected to The World According to Gap, and have responded every time by rushing to escalate the arms race, reducing everyone's safety in the process. We have reached massive and insane levels of "overkill." Overkill, it should be noted, is not a biologically meaningful concept. It has been defined by James Real as "pouring a bucket of gasoline on a baby who is already burning quite nicely."

The modern Neanderthal, it seems, is an intellectual descendant of Procrustes, that unpleasant Greek who adjusted the anatomy of hapless travelers to fit his special iron bed. When confronted with anything new, today's Neanderthal immediately reverts to Procrustean thinking, trying to fit the new problem onto his old conceptual framework. Hence, more nuclear weapons are better than fewer, the nuclear missile becomes just artillery or the catapult writ large, nuclear wars can usefully be threatened, fought, limited, and won, just like conventional wars always were.

Living things like to win, and there is no reason to think that primitive human beings were any different. Like football coaches who claim that a tie is about as satisfying as kissing your sister, our Stone Age cousins did not aim for a draw. Those who did were probably defeated by others who played to win. Evolution tends to be a zero-sum game: if one side wins, the other side often loses proportionately, with the total of all payoffs equal to zero. This is because there is only a limited amount of "niche space" available, and as we have already seen, most populations remain stable over time. So, my gain was your loss, and if you profited, it was ultimately at my expense unless, of course, we were relatives or potential reciprocators. Once more, the ground rules have changed, but the Neanderthal within us still goes by the old version, the one that prescribes victory.

Aggressiveness also raises the question of the adaptive potential of fighting. Animals don't fight all the time, and neither do human beings. We don't *need* to fight in the same way that we need to eat or to sleep. However, throughout most of our evolutionary history, we almost certainly did fight on some occasions, notably over food, living space, or mates. Primitive "warfare," for example, is common among nontechnologic peoples such as the Yanomamo of the upper Amazon or the Tsembaga Maring of highland New Guinea; in such cases, however, the mortality rate is relatively low and "war" is much closer to skirmishing. In addition, success in battle can bring success in life: animal protein, status, *lebensraum*, and often, women.

Today the danger is clear: there is simply no way that nuclear war can be anything but a losing proposition, and yet, we readily respond to frustration, competition, and threat by manning our battle stations, as though the world hasn't changed and warfare might still be worthwhile. No less than other living things, we are masterful cost/benefit strategists, doing something—such as fighting—when its benefits outweigh the costs. As we have seen, Homo sapiens' evaluation of such circumstances was fixed during our long evolutionary childhood, when war worked. Now the cost/benefit equation has changed drastically, and nuclear war can only bring disaster, but we nonetheless find war (any war) strangely attractive, and the prospect of victory, nearly irresistible.

The second aspect of prenuclear mental processes revolves around limitations in our ability to perceive danger. When the Neanderthal felt threatened, he generally was: a charging herd of mastodons, a forest fire, a bellicose fellow Neanderthal. And by the same token, when he felt safe, he generally was. (The masculine pronouns are employed here intentionally because of the high probability that *male* Neanderthals were the major instigators of and respondents to aggression and threat.) But once again, times have changed. As psychiatrist Jerome Frank has pointed out, nuclear weapons have no "psychological reality": they can not be seen, smelled, heard, or felt, and so the Neanderthal feels safe . . . even though he isn't. The threat is there, and is very real, but it lacks the tangible reality of a mugger with a knife, for example, something to which the Neanderthal is well attuned but which is actually far less threatening to us all. Concern about nuclear weapons vanished almost entirely for a time following 1963, when the limited test-ban treaty prohibited weapons testing in the atmosphere (and in space, and underwater). Such testing actually continued, however, and at a more rapid pace than ever, but it was moved underground: out of sight and hence, out of mind for most modern Neanderthals who, like the caricatured ostrich, feel safe so long as they can no longer see the danger.

During the Vietnam War, millions were motivated to end a conflict whose reality was brought home in daily newscasts. By contrast, today's nuclear Neanderthal is generally unmoved by a conflict that hasn't yet begun, and which will be over long before he or she has the opportunity to protest. One of the great challenges facing today's peace movement, then, in addition to overcoming the deep-seated human tendency to misperceive the uses and abuses of aggressiveness in a nuclear age, is a need to overcome the equally deep-seated tendency to restrict our perception of danger to those situations that connoted danger in the past, and which—thanks to the invention of nuclear weapons—are no longer valid hallmarks of risk today.

In the same vein, consider another interesting suggestion from Roger Fisher: because of the bloodless high technology and "nukespeak" surrounding nuclear weapons and their command and control, a president could order their use without really understanding, deep in his guts, what he is doing. Accordingly, perhaps we should replace the computer-coded little black briefcase with a small capsule, implanted next to the heart of a trusted aide. Then, to send forth the Emergency Action Message, the president would have to do something more human and hence *more real* than reciting "Execute SIOP Plan 1-GA4Z, PDQ." Instead, he would be required to cut open the aide's chest, dipping his own hands in human blood. Bright red human blood might be just what's needed, to shock even the most insulated Neanderthal back to reality.

Pain is a useful warning device, telling us that something is wrong. Whether a toothache or a stubbed toe, pain is an unpleasant and hence, effective way of getting our attention. Therefore, we are inclined—almost by definition—to avoid pain. But pain can be emotional as well as physical, and few things are as emotionally painful as facing the danger of nuclear annihilation. "Have a nice day!" proclaim the mindless, yellow smiley-face stickers that became emblematic of the 1970s; thinking about nuclear war will not help you to have a nice day. So the twentieth-century nuclear Neanderthal avoids pain by avoiding the issue. Ironically, once again, a "defense mechanism" that was useful in other contexts helps propel us toward unparalleled catastrophe in the modern age.

Other aspects of our prenuclear mentality further conspire to insulate us from even perceiving the nuclear threat. The very enormity of nuclear war, the literal amounts of power and energy involved, leave us uncomprehending. For the caveman, "hot" is 100 degrees in the shade, or perhaps the temperature of boiling water, or a fire. Maybe molten metal, at the extreme. But not 100 million degrees, the temperature inside a thermonuclear fireball. And what would hundreds of millions of deaths be like? Or a nuclear winter? We are unable to incorporate these realities into our caveman consciousness. It is not just that we don't want to entertain such thoughts, but rather, we are not able to do so: like visitors from another planet, whose antennae are not adjusted to the wavelengths of this new, nuclear world, we wander numb and unresponsive.

For the Neanderthal, as for Ecclesiastes, there was nothing much new under the sun. Even humanity's cultural advance, rapid by biological criteria, must have seemed excruciatingly slow and uneventful to those living through it, day by day. For most of us, even now, the past is a good guide to the future. If it hasn't happened yet, it is generally a good bet that it never will. Accordingly, we quickly reassure ourselves with the observation that nuclear war hasn't happened yet, in the four decades since 1945. Deterrence, we are told, works.

Few of us would take the fact that we are now alive as evidence that our death will never happen; and yet, many of us are reassured that nuclear war will not happen, because it hasn't happened yet. Freud suggested that the human capacity for "denial" is essential if we are to function normally from day to day. After all, our personal death is inevitable, so why obsess about it? Nuclear war, on the other hand, is not inevitable—but neither is it impossible or even unlikely, especially if our biology has its way and those most likely to oppose it are also most likely to ignore the whole issue, leaving the field clear to those military officers, politicians, and industrial contractors who profit personally from the arms race, and hence, are perfectly willing to gratify that aspect of their prenuclear mentalities.

Then there is habituation, the simplest and most primitive of all learning processes: learning *not* to respond. It is clearly maladaptive for an animal to respond to every stimulus that comes its way, and not surprisingly, then, even animals as simpleminded as flatworms stop responding to irrelevant stimuli after awhile. The progressive accumulation of nuclear weapons in the world has certainly not been gradual as measured in evolutionary time, but nonetheless the growth of superpower arsenals has taken an entire generation, a rather long time for quick-moving, conscious creatures. Almost insensibly, we have built more and more, never entirely registering what we were doing and thus, never adequately outraged.

The human nervous system is sensitive to short-term changes in stimulation, but adapts quickly to constancy or gradual change. Hence, we notice (adaptively) a new smell or a sudden sound, but we become literally insensitive to that same smell or sound if it is continued for a period of time without obvious consequence. Since nuclear weapons haven't been used directly against people since 1945, we have become almost as unaware of their presence as we are of the constant drone of a refrigerator motor, or our own smells in the bathroom.

It is said that if a frog is dropped into boiling water, it will jump out and save itself. But if placed in cold water that is then heated gradually, degree by degree, it will never notice the difference and boil to death. Our temperature has been rising, but most of us haven't been noticing.

There was widespread shock in the United States when the *Lusitania* was sunk by German submarines in 1917 and several hundred innocent civilians lost their lives. Then, the Italian use of poison gas in Ethiopia; the German bombing of the Spanish town of Guernica (immortalized by Picasso's painting); the killing of 35,000 Dutch civilians in Rotterdam; 100,000 or more in Dresdan, Hamburg, Tokyo, and finally, Hiroshima. If anything, our outrage has, over time, diminished, and we are no longer shocked by things that once seemed appalling. We have habituated to horrors.

After German Zeppelins had dropped some bombs over London in 1914, George Bernard Shaw wrote to the *London Times* proposing that the London County Council look into constructing air-raid shelters for the city's children, in the event that such attacks become more frequent. The newspaper reproved Shaw editorially, pointing out that it was inconceivable that so civilized a country as Germany—even when at war—would ever stoop to something so barbaric!

Another aspect of prenuclear thinking revolves around patterns of group identification, notably nationalism. The primitive human being relies heavily upon his fellows for success and survival, just like any other social animal, and in fact, more than most. For the many thousands of generations that preceded modern times, our ancestors sought and obtained safety and ultimately, reproductive success, in groups. And for most of that time, the larger the group the better. Families and tribal bands consisted of relatives and/or reciprocating colleagues, whereas by contrast, other families and other tribes were likely to be competitors and often antagonistic as well. Accordingly, there was an evolutionary payoff in what modern sociologists call "in-group amity, out-group enmity," mediated in part, perhaps, by kin selection.

Once again, however, a biologically adaptive tendency seems to have gone awry, leaving us at the mercy of "supernormal releasers," those exaggerated traits to which we are so peculiarly sensitive. Our allegiance is now readily seduced by culturally hyperextended groups known as nations, which offer much of the primitive gut-level satisfaction of family and tribe. Unlike rooting for the Detroit Lions, however, or belonging to the Benevolent and Protective Order of Elks, the modern patriot clings to a social unit that is *not* benign, and which in fact has elevated itself above the individual, threatening him and her with destruction for its own, national, "security."

When we review the annals of human nastiness, we tend to think of murder, rape, torture, arson, or assault, and yet, by far the greatest weight of human evil against fellow human beings has not been committed by the solitary sociopath, but rather, by Homo sapiens acting in a group. It is an excess of devotion, of subordination of the self to the group, not an excess of self-seeking, that leads us most dangerously astray. In human history, more harm has come from obedience than from disobedience (although we are typically warned against the latter, not the former). Similarly, more harm comes from excessively zealous group orientation than from individualism. The nation as a social unit has no fundamental, biological legitimacy, and yet it retains a deep human allegiance, typically because it mimics the primitive Neanderthal yearnings for deeply meaningful group associations.

As Freud pointed out in his *Thoughts for the Times on War and Death*,

> [the] state has forbidden to the individual the practise of wrong-doing, not because it desires to abolish it, but because it desires to monopolize it . . . A belligerent state permits itself every such misdeed, every such act of violence, as would disgrace the individual . . . Well may the citizen of the civilized world . . . stand helpless in a world that has grown strange to him.

Unfortunately, the world in its barbarity has not grown entirely strange to the human beings in whose name we prepare the most hideous atrocities. For so long as these acts are to be perpetrated in the name of "national security," which the nuclear Neanderthal erroneously equates with personal security, eons of evolution conspire to pave the way and legitimize the most illegitimate of plans and justify the most unjustifiable behavior.

Such collective self-deception also leads to "evil empires," and to a self-righteous certainty that whatever We do is good, and whatever They do is bad, even though such actions are often mirror images of each other. For example, when the USSR (a diverse group of Homo sapiens inhabiting eastern Europe and west-central Asia, largely having murderously displaced the indigenous inhabitants) invaded Afghanistan in support of a locally unpopular foreign government threatened by revolution, we denounced that as aggression. But when the United States (a diverse group of Homo sapiens inhabiting North America, largely having murderously displaced the indigenous inhabitants) invaded Vietnam in support of a locally unpopular government threatened by revolution, that was proclaimed a noble cause.

Neither side is able to see the world as the other side sees it. No surprise there; after all, during the 99.999 percent of our evolution that preceded modern times, such a leap of empathy was not needed. It was enough to take care of ourselves, secure in the knowledge that we were right, and they were wrong and in fact not really human anyhow. Precisely that tendency, to dehumanize one's opponents, greased the way for behaviors toward the Other that would never be permissible within the social group. For generations, human beings have defined themselves and their fellow group-members as "human," and considered that non-group-members are so alien as to be literally inhuman. Dehumanizing slang is typically directed toward such aliens, especially when the two sides are at war. Moreover, such attitudes facilitate the declaration of war in the first place. Kin selection may well be operating here, since it is a primitive biological tendency to behave benevolently toward relatives, competitively toward nonrelatives. This vestige of our evolution, adaptive in the distant past, could well facilitate a disregard of our shared humanity which in fact is not "inhuman," but rather, all too human.

Fortunately, dehumanization of the enemy, although perhaps encouraged by biological tendencies, is a delusion. It can readily be dispelled, simply by the recognition of our shared humanity. In *Homage to Catalonia,* his account of the Spanish Civil War, George Orwell described the homely situation that led to such a recognition on his part:

> At this moment a man, presumably carrying a message to an officer, jumped out of the trench and ran along the top of the parapet in full view. He was half-dressed and was holding up his trousers with both hands as he ran. I refrained from shooting at him. It is true that I am a poor shot and unlikely to hit a running man at a hundred yards . . . Still, I did not shoot partly because of that detail about the trousers. I had come here to shoot at "Fascists"; but a man who is holding up his trousers isn't a "Fascist," he is

visibly a fellow creature, similar to yourself, and you don't feel like shooting at him.

Similarly, Western politicians—arguing for additional nuclear weapons as "bargaining chips"—like to point out how dangerous it would be if they had to go into latest round of nuclear negotiations "naked." But perhaps that is precisely how it should be done. Maybe the world's people (in whose name, after all, such negotiations are being conducted) should require that high-level discussions of this sort be carried out only by naked human beings. Devoid of artifice, fully revealed as vulnerable, biological creatures, maybe they would laugh at each other and at their own pretensions; maybe they would find the wisdom and humility to act in the service of life instead of death.

Finally, let us consider another aspect of our prenuclear outlook that impedes antinuclear action. There was a certain primordial, adaptive wisdom by which the Neanderthal apportioned his or her energy. We avoid tackling problems that are too large and whereas we love projects, we abhor those that can't be completed successfully. Thus, we derive our day-to-day satisfactions from events that are on a personal, human scale; family, friends, work, recreation, etc. It simply does not pay to wrestle with a hurricane, a volcano, or an earthquake. At the very least, it is a waste of time to tackle oversized opponents. It may also be dangerous, and through the course of our evolution, natural selection has almost certainly acted against the Don Quixotes among us, who dreamed impossible dreams and broke their lances against unyielding windmills.

Unquestionably, nuclear war is the unyielding, oversized opponent par excellence, not only in the magnitude of its effects, but also in the size of the forces that must be tackled: bureaucratic, military, political, economic, etc. In short, the problem is big and each of us is small, so we readily answer the primitive urgings within, and leave the problem to someone else, to our nuclear priesthood and our parent/leaders, who, after all, are wiser and stronger than us, and whose benevolent attention to such things permits us to go about living our own little lives, reaping whatever primitive, personal-sized, and biologically appropriate rewards we can achieve.

> The hawk that motionless above the hill
> In the pure sky
> Stands like a blackened planet
> Has taught us nothing.

writes Edna St. Vincent Millay, in "The Bobolink."

> seeing him shut his wings and fall
> has taught us nothing at all.
> In the shadow of the hawk we feather our nests.

And so we go on, feathering our nests, oblivious to the shadow of the hawk—or perhaps seeing it but looking the other way because such sights are inconvenient, interfering as they do with the primitive, universal yearning to go on with our little day-to-day lives.

There may also be another factor, one that virtually everyone will deny, but which may nonetheless be worth examining. And this is the deep fascination that some people find with nuclear weapons themselves, precisely because they are so extraordinary, so powerful, and so impressive. They offer to Homo sapiens, a weak-bodied and anatomically unimpressive biological creature, the opportunity to identify with a level of force—and to some, of beauty—that is potently seductive. For example, consider this account, by Brigadier General Thomas Farrell, of the world's first atom bomb test, at Alamogordo, New Mexico:

> The effects could well be called unprecedented, magnificent, beautiful, stupendous and terrifying. No man-made phenomenon of such tremendous power had ever occurred

> before. The lighting effects beggared description. The whole country was lighted by a searing light with the intensity many times that of the midday sun. It was golden, purple, violet, gray and blue. It lighted every peak, crevasse and mountain range with a clarity and beauty that cannot be described but must be seen to be imagined. It was the beauty the great poets dream about but describe most poorly and inadequately. Thirty seconds after the explosion came, first the air blast pressing hard against people and things, to be followed almost immediately by the strong, sustained, awesome roar which warned of doomsday and made us feel that we puny things were blasphemous to dare tamper with the forces heretofore reserved to The Almighty. Words are inadequate tools for the job of acquainting those not present with the physical, mental and psychological effects. It had to be witnessed to be realized.

Thirty-six years later, psychologist Nicholas Humphrey delivered a nationally broadcast lecture on the BBC, titled "Four Minutes to Midnight," in which he gave a different perspective on the human potential of such power and beauty:

> "Do it beautifully!" says Hedda Gabler to Lövborg, as she hands him the gun. Oh yes, we'll do it beautifully. What more beautiful way to do it than in the way that poets dream about, but describe most poorly and inadequately? But the gun goes off by accident, and Lövborg dies miserably, shot not through the heart but through the balls.

The above only briefly evokes the perils of the nuclear Neanderthal as reflected in aggressiveness and success, perceptions of risk, group identification, and finally, impediments to activism. There are many other components as well, such as a tendency to use habitual, noncognitive behavior under stress (after all, it was quick reflexes rather than creative problem solving that saved our ancestors from the crouching saber-tooth); an inclination to imbue leaders with superhuman qualities of wisdom to match their superhuman qualities of power; a tendency to assume the worst in everyone else (paranoia), often employing a reflection of our own unpleasant traits and in the process, generating a self-fulfilling prophecy that produces precisely the effect we most fear; a perverse insistence on organizing the post-Hiroshima world around an intellectual system of primitive threat and punishment (i.e., deterrence), which, although wrapped in arcane intellectual pretensions, is fundamentally one of the simplest, most primitive ways in which one animal coerces another, and among complex animals, one of the least effective.

Once confronted with the threat posed by nuclear weapons—even after the denial, the habituation, the disordered priorities of national strength and national security, of risk, danger, error, and misperception, of enemies and of personal powerlessness are finally penetrated—even then, human beings typically reach for Brahean solutions: for example, seeking to make deterrence "more secure," rather than seeking to abolish nuclear weapons altogether. We prefer to tinker assiduously but superficially, toying with self-gratifying cosmetic adjustments that may help somewhat at the margins, while in fact maintaining our self-perceived interest in keeping things fundamentally as they are and at the same time gratifying our felt need to act "responsibly." Our failure to respond to radical problems with radical solutions is not surprising to anyone familiar with the general history of Homo sapiens, but in the case of nuclear weapons, it is almost exactly equivalent to rearranging the deck chairs on the *Titanic.*

OUR PENCHANT for Brahean solutions is perhaps most dramatically revealed in the latest Star Wars schemes of the 1980s. Here we see a merger of humankind's pervasive faith in technology—the higher "tech" the better—with our penchant for responding to anxiety engendered by nuclear weapons by clutching yet more firmly the weapons themselves. In his book *Technics and Civiliza-*

tion, Lewis Mumford pointed out that "the belief that the social dilemmas created by the machine can be solved merely by inventing more machines is today [1934] a sign of half-baked thinking which verges close to quackery." Writing more than a decade before the first nuclear weapons were used, Mumford certainly did not have Star Wars in mind, but no better description of this very dangerous fantasy has ever been proposed.

Ever since President Reagan's now-famous Star Wars speech of March 1983, in which he proposed the development of a system to render offensive nuclear weapons "impotent and obsolete," growing attention has focused on the technological, economic, and strategic implications of such a plan. Analyses have proliferated as to whether or not Star Wars (or, as the Administration prefers to call it, the Strategic Defense Initiative) is technically feasible, economically affordable, or strategically desirable. Such debate and such concern are appropriate.

However, there has been surprisingly little attention to the psychological appeal of Star Wars. Whatever its merits or demerits technically, economically, or strategically, Star Wars is brilliant, psychologically. It offers a textbook portrait of prenuclear thought processes being applied to the fastest moving of today's hares. Star Wars is psychologically brilliant because it plays right into our primitive yearnings, precisely at a time when the old reliance on nuclear weapons for security is beginning to look increasingly threadbare to millions of people. At the same time, it is phenomenally obtuse because it is itself an example of those primitive yearnings, and at the highest government level.

Star Wars supporters claim that it would be morally superior to deterrence, since it will defend people, rather than threaten an opponent, or if necessary avenge oneself. In this respect, it is interesting that nuclear hawks, many of whom have made a career out of defending the use of planetary threats—that is, deterrence—are suddenly rushing to attack it, while nuclear doves, for many of whom deterrence is profoundly distasteful, find themselves defending it.

It is not difficult to see why some people support Star Wars:

high-level officials eager to remain in the good graces of their boss, military officers whose careers have suddenly become hitched to its research, testing, and eventual deployment, and high-tech military contractors who can smell a lucrative contract miles away. ("Dollars from heaven," gushed the *Wall Street Journal.*) There is something primitive, and understandable about that old motivator, greed. Among others, there is the appeal to some other deep-seated caveman inclinations, notably the feeling that security can be achieved by more weapons, certainly not by fewer. The attractiveness is an old one, the hope of having one's cake and eating it too: having one's missiles and hiding from them too. Or we might call it the Humpty-Dumpty syndrome: by the very nature of nuclear weapons, our national security has taken a great fall, and now the king is saying that he can *too* put it together again, if we only give him enough horses and enough men.

We must also not forget the seductiveness of high-tech itself, as Mumford warned. Despite ecological consciousness, small-is-beautiful, and so forth, it remains true that many of us have been mesmerized by a continuing, naive faith in technology, and a hope that science will save us, that machines will save us from machines. Pie-in-the-sky may be old hat, but armor-in-the-sky, a genuine astrodome from sea to shining sea is a potent lure, especially because it excuses us from dealing with the really difficult but ultimately unavoidable problem of working out negotiated, political settlements with our adversaries. After all, our technological skills are generally superior to our social skills, so when in doubt, we turn with relief to technology.

And finally, there is the appeal of unilateralism. After all, even ardent Cold Warriors are becoming jaded about somehow "beating" the Russians in the accumulation of offensive overkill. Star Wars offers them a new Holy Grail, an arena of strategic competition that plays to America's strength—technology—with the prospect that we'll just go ahead and solve this problem like real men . . . by ourselves.

Behind it all, it may be that the most cogent psychological prop for Star Wars is a cynical high-level recognition that it will guarantee an indefinite arms race, since as long as they are threatened with the prospect of a potential American Star Wars system, the Soviets will never agree to arms limitation, never mind reductions.

"MOM, can I have my birthday party early this year?" Nothing very unusual in this, coming from an impatient and precocious five-year-old. A bit more surprising, however, was the reason for the request: " 'cause I don't want to miss mine if there's a nuclear war."

A colicky baby, or a teenager with a drug problem demands our attention. A child with leukemia, anorexia, or even a bad case of acne quickly mobilizes parental concern and assistance. The alienated child, who tunes out of adult society and turns on to acid rock and drugs, is no less frightening or motivating for most parents. Some of this alienation has doubtless occurred as long as there were mammals and their awkward adolescent offspring, just beginning to surge with hormones and to seek their place in the social group. But as if this wasn't troublesome enough, the last half of the twentieth century has added something new: fear of nuclear war, that is, not so much fear that the future is uncertain, or likely to be difficult or even perhaps unpleasant, but rather, fear that there may not be a future at all.

Since the dawn of human consciousness, we have struggled with the reality of our own demise. It comes with the territory, in a sense, one of the bitter fruits of awareness. And yet, it is still difficult enough to talk with our own kids about what Kurt Vonnegut calls "plain old death." No wonder it is even harder to confront nuclear war, with its image of utter annihilation. Many parents cannot even talk with their children about sex, perhaps in part because of a desire for sexual privacy, discussed earlier. However difficult the facts of life, many more people shrink from discussing nuclear war, the facts of death. And yet, just as children simply cannot grow up without encountering sex, they also cannot grow up today without encountering nuclear war, in thought if not reality.

Most parents are frustrated, bewildered, shaken, and confused when faced with their children's anxieties about nuclear war. After all, we pride ourselves in "taking care" of our children, and nuclear war—more than anything else—confronts us with the harsh reality that our web of parentally provided security is really an illusion. We can invest in piano lessons and computer camps, and insist on daily dental flossing, so as to be good parents, and yet, if the bombs go off we have failed, utterly. Moreover, the fear and estrangement in some children is such that even if there isn't a nuclear war, we have failed anyhow.

The simple fact is that our children *are* threatened, as never before, and they know it. Never before have young members of the species Homo sapiens had serious reason to doubt the continuation of their species. Never before have we had a mushroom-clouded future. The "doomsday clock" of the *Bulletin of the Atomic Scientists* has been moved ahead to just three minutes before midnight, and responsible experts warn increasingly of the dwindling likelihood that we shall avoid nuclear war before the next century. To some extent, the current nuclear peace movement is responsible for the epidemic of nuclear anxiety among our youth; that is, our children are worrying because they are being told that they have something to worry about. The reality is, however, that blaming the peace movement for the widespread alienation and nuclear fear is like blaming the person who shouts "Fire," or who turns in the alarm, rather than the arsonist. The challenge for today's parents is to help their children respond effectively to that alarm, not to ignore, deny, or cover it up.

We may never know the precise psychological cost of living under the Bomb of Damocles, but it seems likely that an earlier generation has paid its share. Thus, many of today's parents are themselves veterans of "duck and cover" drills of the 1950s, and the strontium-90 still in their bones mixes all too well with a long-suppressed anxiousness in their hearts. Meanwhile, the current generation is not exempt. Just as the danger of nuclear war is probably higher now than at any time since the Cuban Missile Crisis, nuclear anxiety almost certainly has been keeping pace. A recent study by psychiatrists William Beardslee and John Mack has shown that one half of American children become aware of nuclear issues before age twelve. Among older children, one half thought this awareness affected their plans for marriage and the future. The Beardslee and Mack study, based on interviews with hundreds of school-age children, shows that children are "deeply disturbed" about the threat of nuclear war, profoundly pessimistic, and often, just plain scared.

What's wrong with being scared? According to psychoanalyst Sybille Escalona,

> profound uncertainty about whether or not mankind has a foreseeable future exerts a corrosive and malignant influence upon important development processes in normal and well-functioning children.

After all, "Young people come to terms with the adult world as long as it holds out a reasonable promise for fulfillment in some spheres of living." For many of our children, that "promise" looks more and more like a lie, and accordingly, children may be having more and more difficulty coming to terms with the adult world. A recent study of teenage California "stoners"—who had viewed the murdered body of one of their friends, yet been unmoved by the experience—revealed that these youngsters all felt hopeless about their own future, as though they were being denied any "promise for fulfillment" by the threat of nuclear annihilation.

The problem, however, is not simply how adults can best assuage the fear of our children, how we can minister to their worries, calm their anxieties, and help them to overcome their growing nuclear fear. The issue is not simply a need for juvenile reassurance, any more than a child caught in a burning house needs reassurance; reassurance be damned, the need is for rescuing. The problem is real, not imaginary; one not of attitude but of situation. The first step for parents wanting to help diminish their children's nuclear fear is therefore to recognize that the fear is legitimate—not only because it is real within them, and hence partaking of the "reality" of a psychosomatic complaint—but also because it is based on a real threat outside of them.

It is said that before shipping their children off to Babi Yar, thousands of Polish parents were very careful to see that they had brushed their teeth, combed their hair, and buttoned up their overcoats. Such fastidiousness may have made the parents feel good at the time, but cannot (in retrospect, admittedly) qualify as "good parenting." Similarly, good parenting in the Nuclear Age involves more than simply ministering to the fears of children, as though the fear itself is the problem. It must also seek to diminish the causes of such fear. That is, good parents in the 1980s will not only try to make their children *feel secure*, but also help them to *be safe*. Fortunately, doing one may well be the best way of doing the other. And in the process, curing some alienation in and between both generations.

Victorian sexual taboos did not make sex go away, just as parental reluctance to talk about it does not make it easier for children to become fully sexual adults. And just as it is especially difficult for parents to talk about sex unless they are both intellectually and emotionally comfortable with it themselves, parents will likely have difficulty talking with their children about nuclear war unless and until they have informed themselves about the issues.

In their report on children's attitudes toward nuclear war, Beardslee and Mack note that

> At each stage of development, the child mitigates disappointments by looking ahead and building a vision of the future in which he or she may possess what cannot now be had, or in which it is possible to become what he or she is

incapable of being now. A healthy ego ideal builds out of possible goals or standards that are both realizable and worth struggling to achieve. But the building of such values, or of an ego ideal, depends on a present life that is perceived as stable and enduring and a future upon which the adolescent can, at least to some degree, rely.

The two psychiatrists go on to ask:

> But what happens to the ego ideal if society and its leaders are perceived cynically and the future itself is uncertain? Furthermore, how does it affect the ego ideal when the reason for that uncertainty is readily perceived to be the folly or "stupidity" of the adults around the adolescent who, because of perceived incompetence, greed, aggressiveness, lust for power, or ineffectualness can leave their children no future other than a planet contaminated by radiation and on the verge of incineration through the holocaust of nuclear war. In such a world, planning seems pointless, and ordinary values and ideals appear naive. In such a context, impulsivity, a value system of "get it now," the hyperstimulation of drugs, and the proliferation of apocalyptic cults that try to revive the idea of an afterlife while extinguishing individuality or discriminating perception, seem to be natural developments.

The best antidote for nuclear despair and alienation is not group therapy or chic attempts to "get the fear out," but rather, activism; this applies to children no less than adults. Moreover, just as children in the 1980s seem otherwise likely to grow up (if they do so at all) with a heavy dose of hopelessness and anger toward the adults who have created their world and now seem relatively uninvolved in seeking to improve it, children can profit enormously from images of their parents as active, courageous, concerned, and powerful people. Children are strongly influenced by the adult role models around them. Sybille Escalona notes that "growing up in a social environment that tolerates and ignores the risk of total destruction through voluntary human action tends to foster those patterns of personality functioning that can lead to a sense of powerlessness and cynical resignation."

In addition to the intrapsychic and personality costs, there may also be long-term social costs, borne ultimately by a society that will someday look to these children for leadership: "Growing up with the full knowledge that there may be no future and that the adult world seems to be unable to combat the threat can render the next generation less well equipped to avert actual catastrophe than it would be if the same threat existed in a different social climate."

On the other hand, growing up in an environment that fosters a loving, caring, and activist orientation toward all that is beautiful and special about this planet can encourage those patterns of personality functioning that lead to a sense of joy, power, and self-worth.

Let us also remember that we and our children are in this together, for better or worse. It is, in fact, the ultimate in togetherness: sharing the same fate on a small and endangered planet.

There have not as yet been any studies of children who grow up with an activist orientation toward nuclear war, but such children will probably not become stoners, nor will they feel helpless, hopeless, or despairing. They may, however, feel quite alienated from "the system" or a government that embraces nuclear weapons, and rightly so. But they will almost certainly feel connected to their family and to the human race. They may even develop a renewed faith in democracy, or at least, a renewed hope for its potential. Erik Erikson has suggested that infancy establishes a foundation of hope and "basic trust," followed by childhood and its rudimentary training in will, purpose, initiative, and skill, and then adolescence with its basic grounding in "some system of fidelity." And the greatest fidelity, perhaps, is not to an imagined god, or political system, a spouse, or even to oneself, but rather, fidelity to the planet.

It is ironic indeed that some parents are more upset that their children are upset about nuclear war, than about nuclear war itself. Many people, of course, do not accept personal responsibility for our nation's military and geopolitical policy. Insofar as that is so, the nation's failure to provide a safe world for them and their children is often less troublesome for them personally, since it is not seen as a personal failure on *their* part. But most parents, by contrast, do accept responsibility for the psychic wholeness of their children, not to mention their physical safety; accordingly, many adults have become active in the nuclear peace movement because of the despair and fear they have seen in their children (not to mention their own fear *for* those children.) These parents recognize implicitly the failure of the parents of Babi Yar.

What they may not realize, however, is that by becoming active themselves, they are almost certainly helping their children psychologically as well. To a child, parents are immensely powerful, and almost by definition, successful at what they attempt. So the active, committed parent, alienated from the nuclear nation-state but deeply connected to life itself, may be not only the best long-term hope for a child, but also the best short-term confidence builder. In a recent discussion among 17 students in a second grade class, 16 said they worried about nuclear war. Only one had hope for the future. He explained why: "I know there won't be a nuclear war because my daddy goes to meetings all the time to prevent it."

AS MODERN primitives, we are influenced by our evolutionary past. But this is not to say that we are prisoners of that past; indeed, we are in thrall only so long as we remain unaware of that influence. Once we recognize the outmoded source of our inclinations, the vestigial Neanderthal mentality like a swollen appendix now threatening to burst, we become empowered to excise it.

We are, after all, the most adaptable creatures on earth. We have given up slavery, the divine right of kings, human sacrifice, and dueling, all of which were at one time considered indelible reflections of "human nature." We can learn all sorts of things, like languages, music, or even respect for the rights of others, and we can inhibit inclinations that we recognize as inappropriate, like our muscle-headed Neanderthalism and reflex resort to Procrustean thinking. When it comes to the Neanderthal mentality and nuclear weapons, the conflict between biology and culture in human affairs is more pronounced and more dangerous than in any other realm. And yet, there is hope. Having recognized a problem and correctly identified a threat, Homo can be wonderfully sapient, correcting dangerous situations even if this includes correcting himself.

As Sigmund Freud saw it, culture has a responsibility to win back humanity from the ascendancy of our unleashed instincts:

> The fateful question for the human species is whether and to what extent their cultural development will succeed in mastering the disturbance of their communal life by the human instinct of aggression and self-destruction. It may be that in this respect precisely the present time deserves a special interest. Men have gained control over the forces of nature to such an extent that with their help they would have no difficulty in exterminating one another to the last man.
>
> —*Civilization and Its Discontents* (1930)

More than fifty years later, the problem is if anything more acute than Freud described, and somewhat different as well: our human "instincts" may well be the problem, but so is culture. Thus, it is precisely the runaway elaboration of nuclear culture, and its uncoupling from basic biological wisdom and inclinations, that so endangers the entire planet today. But Freud was also on target: We cannot wait for biology to save the day. Biological evolution is simply too slow. The responsibility cannot be foisted off on natural selection, any more than on government leaders. No, the responsibility is ours. Cultural evolution, acting in recent historical time, caused our nuclear mess. What is needed today, as Gunnar Myrdal put it, is "not the courage of illusory optimism but the courage of almost desperation."

According to Greek mythology, the gods punished Prometheus—who had impudently given fire to human beings—by chaining him to a great mountain, whereupon he was visited daily by a vulture, who chewed on his liver. Modern human beings, biological creatures acting not in deliberate evolutionary time but in a cultural frenzy, have unleashed a much more dangerous fire than Prometheus could ever have imagined. And this fire is made all the more lethal by the fact that, deep inside, we really aren't very "modern" ourselves. In *Prometheus Bound*, Aeschylus asks:

> Prometheus,
> Prometheus, hanging upon Caucasus,
> Look upon the visage
> Of yonder vulture:
> Is it not thy face,
> Prometheus?

Two thousand years later, Pogo said it more simply: "We has met the enemy and it is us."

CHAPTER NINE

POPULATION

Psychotic Rats, Open Faucets, and Closed Minds

Europe is over-populated, the world will soon be in the same condition, and if the self-reproduction of man is not "rationalized," as its labor is beginning to be, we shall have war. In no other matter is it so dangerous to rely upon instinct. Antique mythology realized this when it coupled the goddess of love with the god of war.

—Henri Bergson (1935)

Living things love to reproduce, not only figuratively but literally as well. That is, love is a means of reproducing, and human beings, no less than other animals and plants, tend to be very good at it. A sexually reproducing species need only leave two surviving offspring for each breeding pair for its population to remain constant from one generation to the next. This is more or less what usually happens, not because the individuals in question are especially concerned about the fate of their species, but rather, because competition among individuals and between species only rarely gives a pronounced advantage to anyone. If various factors in the environment did not act to reduce its numbers, each species would generate a population explosion of unimaginable proportions. If just two adults of any plant or animal reproduce unimpeded during their lifetime, and their offspring follow suit, then after a million years (a very short time in evolutionary terms) the entire earth and in fact, all of the visible universe would be packed with the quivering, living substance of our hypothetical organism. It should cause no surprise, therefore, to learn that living things do not achieve their full reproductive capacity. Their numbers are constantly depleted, both as adults and more commonly as immatures, embryos, or simple sperm and eggs. Some are depleted more than others; this, of course, is the stuff of natural selection.

In discussing natural selection, we concentrated on the survivors, and on those characteristics that confer genetic survival value. Now we must look at the losers. Many factors, separately or in combination, can cause living things to die or fail to reproduce. They can run out of food, or some necessary vitamins or minerals. They can fail to find a mate or a suitable place to live or raise young. They can succumb to the vagaries of climate, such as wind, rain, sun, or drought. They can fall to disease caused by bacteria, protozoa, or virus. They can be eaten while alive (by parasites such as the tapeworm or botfly) or first be killed and then eaten (by predators such as hawk or wolf or human being). Although these are hard fates for the individual victim, they also have the unintended benefit of preventing uncontrolled and potentially catastrophic population increases.

One famous case of misguided human altruism illustrates the importance of natural mortality in maintaining a healthy population. The Kaibab Plateau is an enormous wilderness in north-central Arizona. It supported a reasonably large deer population and a goodly number of predators, all in healthy balance with the environment. But the kind-hearted, deer-loving public, especially the hunters, wanted more deer and felt that this could be accomplished by killing their "enemies."

So, a systematic program of slaughter was initiated. From 1907 to 1923, 11 wolves, 600 cougars, and 3000 coyotes were removed from the Plateau. Before this, the deer population had been about 4000 healthy individuals, their numbers kept in check by the predators. With this check removed, the deer population increased spectacularly, to about 100,000. But long before that number was reached, something was clearly wrong. The deer were obviously running out of food, looking thin and sickly. Trees were being killed by overbrowsing and all vegetation had been stripped bare up to a height of about eight feet, the maximum that a starving deer can reach, standing on its hind legs. A great famine began in 1925 and more than half the deer starved to death in the next two years. By 1940, starvation was still killing more deer than the predators ever had, and the population was down to about 10,000.

More serious yet, the range itself had been gravely injured. Before predator removal it could probably have supported about 30,000 deer, but after the trappers and hunters had done their work, the extreme pressures of an unnaturally high population had greatly reduced the Plateau's ecological vigor. The Kaibab deer were victims of their own uncontrolled population. They not only suffer greatly from starvation, but seriously reduced the capacity of their environment to sustain life in the future.

The story of the Kaibab Plateau deer is in many ways an allegory of the planet earth and modern Homo sapiens. Our history has been one of progressively increasing numbers, first very gradually for thousands if not millions of years, then building up to a mighty roar in the last century. The worldwide human population at the discovery of agriculture, perhaps 10,000 years ago, was about 5 million. By the birth of Christ, there were about 200 million of us. We didn't reach our first billion until around 1850—approximately 52,000 years after Homo sapiens arrived on the scene. Our second billion appeared by around 1930, just eighty years later, and our third billion arrived around 1960, a mere thirty years after that. Billion number four was reached by 1975.

It should be clear that not only have our numbers been increasing at a phenomenal rate, but, even more frightening, the *rate of increase* has been increasing in many countries. Thus, even the phrase "population explosion," often used by people who are alarmed about our numbers, is inadequate. In an explosion, things fly very rapidly out from the source, but with decreasing speed farther from the epicenter, and as time goes on. There is no English word that describes the human population "explosion," an unprecedented event in which both the numbers and the rate have been increasing over time. It is more like an avalanche.

Like all animals, human beings are capable of a very high rate of reproduction, theoretically about twenty children per couple, but in practice more often six to ten in cases of unrestrained childbearing. This high birth rate was adaptive for our ancestors, among whom mortality was high. Because many children would normally die of disease, famine, exposure, or predation, primitive human beings required large families simply to keep even, and for individual parents, the cost of additional children was generally less than the benefit. As the Red Queen told Alice, it was necessary for us to run, just to remain where we were. To get anywhere, we had to go faster yet. To guarantee at least two surviving offspring, on the average, it was necessary to have six or seven, and among societies in which special value was ascribed to one sex (typically male) it was also necessary to have many children to ensure that at least one of the survivors would be a boy.

The explosive increase in human population has paralleled our explosive cultural evolution, in three major successive stages. The first probably occurred at the dawn of human evolution and corresponded to our initial invention of rudimentary culture (language, fire, etc.), which among its various effects also enabled greater survival for our offspring. The second, about 10,000 years ago, coincided with the invention of agriculture. This was a cultural advance of fundamental significance, transforming the bulk of humankind from a hunter-gatherer to an agrarian economy. From dependence on nature's whims, we achieved substantial control over our nourishment. Agriculture gave us predictable food supplies and made possible the concentration of population into cities, where the surpluses allowed individuals to specialize in other, nonagricultural pursuits. The population zoomed.

The third and greatest spurt began with the scientific revolution of the sixteenth and seventeenth centuries and has continued up to the present, having been fueled by the Industrial Revolution and more recently, the great hygienic and medical advances of the last 100 years. The human population is currently on a collision course with our ability to maintain social harmony in our overcrowded world, and with the basic physical and biological resources of the earth.

The Kaibab deer experienced their drastic population increase because their most important source of mortality, their predators, had been eliminated. Their reproductive capacity had always been high, geared to an equally high death rate. Like in an automobile that had always been driven with the brake on, the accelerator was also floored, just to keep the car moving. With the brake suddenly released and the gas pedal still pressed to the floor boards, the vehicle went out of control and crashed.

We have a high natural capacity for reproduction, also geared to a primitively high death rate. In human history, the most important natural braking agents have probably been starvation and disease rather than predation. When our ancestors began perfecting culture, becoming more efficient hunters, gatherers, and food preparers, they greatly reduced their losses from starvation. The invention of agriculture had a similar effect, while also permitting human survival in previously uninhabitable or difficult areas. In the last 100 years phenomenal strides have been made in our long battle against disease. Just as with the Kaibab deer, we are eliminating our cougars and wolves, only for us they are spelled malaria, cholera, and typhus. And just as with the Kaibab deer, our numbers are now skyrocketing; but with life just as with gravity, what goes up has a habit of coming down.

Population biologist Paul Ehrlich has emphasized that the Western world in particular has been practicing and exporting death control. On balance, human mortality has been greatly reduced, especially in the Third World, where it had always been appallingly high. There is, of course, nothing wrong with this; it is laudable. Starvation and disease are still two of our ugliest enemies and eliminating them must rank among the noblest human endeavors. But because the human population has increased beyond the earth's ability to provide consistently and safely for Homo sapiens, we witness the periodic, devastating famines such as those wracking Africa in the 1980s. Moreover, the long-range prospect for supporting an artificially inflated human population is bleak at best, and perhaps utterly impossible. To substitute one form of death for another, resulting from ultimate overpopulation and therefore likely to affect an even larger number of people, is irresponsible. The alternative, fortunately, is simple enough: continue our humanitarian efforts toward death control (in fact, increase them) but also simultaneously export *birth control.* As we ease up on the brake, our only hope for a smooth ride is to gradually release the accelerator as well.

We are faced with a social problem that would not occur if we weren't cultural as well as biological creatures. Thus, while our culture provides death control, our biology is still tuned by evolution to the production of a large number of offspring, in anticipation of the high mortality which is now largely averted. Under conditions of natural selection, analogous changes in mortality might eventually be reflected in an altered reproductive pattern. But this would take time on an evolutionary scale—thousands of years at least—whereas drastic cultural developments are occurring in decades, or less. As with the problem of human aggression, our population problem was born when cultural evolution left biological evolution far behind, and it must now be resolved by cultural forces acting alone. We simply can't wait for our genes to do it.

Fortunately, birth control might actually cost less than the present technology of death control while ultimately being no less important. A successful worldwide birth control program would have to overcome the opposition of certain religious groups as well as the rather paranoid (but sometimes justified) suspicions of certain ethnic groups—notably, many American blacks—who see birth control as a clever guise for racial genocide. Furthermore, it may have to buck a genetically motivated human desire for large families, originally instilled by natural selection long before our culture discovered death control. But when it comes to behavior, our biology is frequently subordinate to our culture, and fortunately it appears that strong culturally induced motivations for limiting family size can overcome any such biological tendency. Cultural evolution in this case offers not only the problem but also the solution. Birth control pills, condoms, IUDs, diaphragms, and sterilization surgery are all the products of human culture, which—if we can summon the requisite wisdom—permit us to thumb our collective noses at our genes.

The liberation of human sexuality from its purely reproductive function, discussed earlier, should facilitate use of contraception. In addition, as economic development proceeds, having too many children becomes an increasing social and financial burden for individual families. A hopeful and consistent phenomenon in recent years has been a tendency for birth rates to decline when socioeconomic conditions have improved. This is a purely cultural happening, the so-called "demographic transition," and it appears to be based on enlightened self-interest. A worldwide effort under the auspices of UNICEF has begun making significant progress in controlling juvenile diarrhea, the number one killer of young children. A double benefit seems to result: not only is suffering reduced immediately, but with their children no longer dying so readily, adults are more inclined to use birth control, thereby reducing suffering in the long term as well.

A similar shift often occurs among animals, when their situation changes such that parents are more successful producing a relatively small number of offspring and investing more heavily in each one, rather than producing a large number of offspring, and giving each one only a very small headstart. In peasant or hunter-gatherer societies, additional children are additional farmers, hunter-gatherers, and/or warriors. And when they die—as they often do—they are replaced. Among technologically advanced societies, by contrast, additional children represent a substantial drain on family resources, since each one requires heavy investments by way of medical care, clothing, and education, while there is relatively little immediate domestic productivity that they can contribute.

If our culture fails to respond to the threat of overpopulation, or if that response is inadequate, our biology just may do the job for us. The populations of many animals are limited by factors that ecologists call "density-dependent." This simply means that as the population increases, a higher number—or more effective yet, a higher percentage—of individuals die. It's a bit like the graduated income tax, with population density substituting for annual income and mortality substituting for the tax liability. Under density-dependent control, mortality increases as the population density increases and it decreases as the density declines. Some species, however, such as grasshoppers, appear to be limited by "density-independent" factors: death rate in such cases varies without regard to the actual population size. Thus, grasshopper numbers may be determined by environmental factors like drought or rainstorms that kill a relatively constant number of individuals, regardless of the total population. But it is the density-*dependent* species that are of greatest interest to us. These are the living things that seem to regulate their own population size. Analogous to a progressive tax structure, species of this sort experience a progressive death structure: "wealthier" (more abundant) populations suffer a proportionately higher mortality.

Density-dependent systems are examples of what engineers call "negative-feedback loops." The idea is really very simple. A positive feedback (the opposite) is what we colloquially call a "vicious circle," a situation in which some deviation from the norm results in a further deviation, with that deviation causing one larger yet, and so on. "The rich get richer" is an example of positive feedback ("and the poor get children"). A nuclear chain reaction is another example. Negative feedback, however, is more intellectually appealing, involving as it does a wonderfully balanced system, preset to absorb deviations and promptly return the system to an acceptable level. For a mundane and frequently overlooked example, consider the home thermostat. It may be set for a certain range of temperature, say sixty-five to sixty-seven degrees: if the house gets colder than sixty-five, the furnace turns on and warms it up to the acceptable range. Should the temperature exceed that range, the furnace shuts off, allowing the house to cool down. A really sophisticated negative feedback system might also incorporate central air conditioning, which would be activated if the temperature got too high.

Biological systems possess a vast array of negative feedback mechanisms, all acting to guarantee the precise physical and chemical conditions needed to maintain life. For example, our blood and body fluids must be kept within extremely narrow limits of acid and alkali if we are to avoid dying in convulsions. A more obvious example would be the maintenance of internal temperatures by a series of mechanisms similar to a household thermostat. Because we are warm-blooded animals, a change in body temperature of more than a few degrees in either direction could mean a coma and death. We unconsciously raise our temperature by shutting off the small blood vessels that lead to the skin, thus preventing heat loss, while rapidly contracting our muscles—shivering—to generate heat. If on the other hand we are getting overheated, more blood is pumped to the skin, where excess warmth is radiated into the air. At the same time, we also exude water onto our body surface—perspiration—which evaporates and cools us even more.

Analogous negative feedback processes operate in density-dependent control of animal populations. Flour beetles thrive in a mixture of dried cereals and grains, as the wholesale grocer is well aware. Left untouched in a feed bin, a small number will multiply rapidly. Eventually, the population will stop growing, in part because their own accumulating wastes are harmful to the animals; as the numbers increase the amount of waste increases likewise. There is always the outside possibility that something similar could occur among Homo sapiens. Thus, as toxic products such as mercury, lead, and PCBs accumulate in our environment, they could eventually interrupt our reproductive processes. And the industrial disaster in Bhopal, India, where thousands died, shows that even without war or widespread ecological collapse, modern civilization has some appalling "downside risks."

Aside from doing themselves in via their own toxic byproducts, flour beetles also kill each other directly. They can be cannibals, eating the young larval mealworms when they get the chance. Normally, in a sparse population, they don't get that chance very often; but as the population grows, individuals are more and more likely to bump into each other, with lethal consequences for the juveniles.

Something similar happens among grizzly bears. Male grizzlies, in particular, kill cubs. This is why the females chase them away before the birth of their young and why a sow bear with cubs is so dangerous to humans. She is highly protective of her young and evolution has told her that they are in danger. When the grizzly population is low, males only rarely get a chance to destroy many cubs, but should the population become inordinately high, we might expect this density-dependent mechanism to take its toll of the young bears, which in turn would restrict further population growth. It seems unlikely, by the way, that male grizzlies—any more than adult flour beetles—are consciously seeking to reduce the local population. Rather, they are probably just taking personal advantage of whatever opportunities come their way. To some extent they may also be following the dictates of selection, which would reward individuals who eliminate potential competitors, despite the fact that in some cases they may be killing their own genes. The overall population effect of such behavior is probably incidental, although no less real as a result.

Both these examples portray density-dependent population control in which adults attacked young. For a final example, in which adults attack other adults, let us journey to the cold waters at the bottom of the North Sea, and look briefly at the lives of the hard-shelled crabs that crawl about down there. Each animal is normally protected from its fellows by a strong external armor. But periodically it must molt, leaving the soft body dangerously exposed. If the sea-bottom population density is sufficiently low, a freshly molted crab will survive its brief susceptible condition. But if the animals are crowded, soft-bodied individuals will be found by their more numerous and invulnerable colleagues, and many will be eaten. High local populations thus tend to decrease.

It is unlikely that such mechanisms have ever operated to control human populations, and certainly no one predicts that cannibalism will exert any real effect on human numbers in the future. On the other hand, aggressive violence between people may be at least partially a function of population pressures. If higher population resulted in more violence, which in turn increased the death rate, we would have a straightforward density-dependent system that could (at least in theory) produce a relatively constant population size. There are several possible pathways linking population density and aggression.

Aggression is more likely to be expressed in a dense population, in which people are constantly bumping into the potential objects of their aggression. Any possible territorial instinct, however unlikely, would be severely stressed by increased crowding. Population density also seems likely to make social harmony more elusive than ever. It is probably no coincidence that our largest urban area, New York City, is generally acknowledged to be ungovernable. And as we have seen, the breakdown of social order has often been cited as a cause of aggression. It is significant that statistics for violent crime tend to be highest where population is the most dense. Furthermore, high populations tend to exaggerate demands for public services, while accentuating the discrepancies between rich and poor. Frustration rises, and with it once more, the tendency for aggression.

Despite these apparently tidy correlations, there is still no hard evidence for an aggressively mediated density-dependent control of human population and even if it was an imminent possibility, it wouldn't be desirable. In fact, none of the density-dependent control mechanisms inculcated in different species by biological evolution can hold a candle to the possibilities of cultural control, because such control would ideally involve *preventing* the problem, rather than dealing violently with it.

Although ecologists are divided as to the full significance of density-dependent controls in nature, they are generally agreed that to some extent at least, the size of a population influences the number that die, even when the lethal factor seems to be independent of density. This also applies to human populations. For example, if a drought or some other vagary of climate restricts the amount of food so that only 50 deer can survive on an isolated island, and 51 are present, one unfortunate must die. Similarly, if there are 200 deer then 150 will starve. This is density-dependent in a sense, in that the greater the population, the greater the number of individuals that die. The human parallel is appalling: if during conditions of periodic famine, a nation can support only 50 million people, but has been artificially inflated to a population of 200 million through massive death control without commensurate birth control, we can expect 150 million deaths when push ultimately comes to shove. And the larger the population, the more people affected.

It is also possible—and in some cases, more likely—that the costs will simply be spread throughout the population, with the survivors all tightening their belts somewhat. Everything we know about human behavior suggests that to some extent this will happen; on the other hand, everything we know about human behavior also suggests that sacrifices imposed by resource scarcity will not simply be distributed equally.

For many animals the environment dictates that only a limited number of individuals will be successful. In a territorial species requiring, for example, two acres per breeding pair, a ten-acre plot can accommodate only five pairs. The surplus, if any, is out of luck. Bobwhite quail generally have a hard time getting through the winter, but those that can obtain adequate shelter usually survive. Since there are usually more birds each fall than their environments can safely accommodate, a certain excess can be expected to die off each year. The environment sets the numbers that can survive and any excess usually perishes. The larger the population, the more animals are forced into suboptimal habitats and the higher the rate of mortality. When such displaced individuals eventually die, it is frequently "because" of predators: in the case of bobwhite quail, great horned owls or foxes. The predators are therefore often blamed for the mortality, whereas actually it is the prey species's environment, relative to its population size, that determines how many will die each year.

Sadly, a similar process occurs in human beings, although minus the predators. In the early 1970s, a typhoon in East Pakistan (now Bangladesh) claimed perhaps *500,000* lives. This terrible human loss may have seemed at the time to be a direct result of natural climatic factors, not in itself related to population size. But in fact, the Bengalis were victims of their own population density: the treacherous Gangetic Delta should never have been populated by so many people and indeed, would not have been if the whole area was not so horribly overcrowded. Like excess quail in winter, doomed to an inadequate habitat, millions of people were forced onto treacherous terrain by the very existence of their compatriots. Similarly with drought-stricken Africa, in which the population density, while less than that of Bangladesh, is nonetheless too high for the resources available.

Among most animals, competition tends to cull out the genetically less fit—since the "better" individuals can be expected to secure a safe place for breeding and survival—just as predators are more likely to eliminate the old and the sickly. But among humans, with our effective and ever increasing cultural buffering, there is virtually no chance for such selection to take place. Given the enormous potential present in every human being, and the equal enormity of personal pain and sorrow, only a gene-obsessed ghoul can take comfort from the notion that natural selection is somehow improving the species by such disasters. But even if such changes were theoretically possible, and even disregarding outraged ethics and morality, human living arrangements have virtually made such efforts biologically impossible: Imagine a child with fortuitously excellent gene combinations, born to an impoverished peasant family where she is one of nine children. There is virtually no real opportunity for this child to escape the perilous life on the Delta or on the arid Ethiopian plateau to which population pressures have driven her family. When the flood comes or the famine descends, she will be swept away, part of a human surplus, at the mercy of nature and an awful kind of density-dependence. How many heirs to Tagore were lost in that typhoon? How many Kenyattas have already starved?

Among many animals, numbers are also restricted by disease and parasites. Our peculiarly anthropocentric viewpoint often overlooks the fact that microbial pathogens are not a problem for our species alone. They are often the major factors controlling population size, especially among most free-living primates, for example. The effect here is once again somewhat density-dependent, since contagion is more easily spread in a dense population than in a sparse one, and overcrowded, malnourished individuals are more susceptible to disease.

If you were a parasite or disease-causing organism, death of your host would be a serious affair. You and your offspring would die along with your former benefactor unless there were opportunity to infect others. Disease outbreaks are therefore most likely when the susceptible population is large. The infection "runs its course" until the remaining individuals are those with generally higher resistance, and the population has thinned out sufficiently so that they are unlikely to contaminate others. There is a strong correlation between local population density in human beings and the chances of contracting disease. Who hasn't been exposed to the annoying cough and sneeze in a crowded bus or theater? And all parents know that their children are more likely to get sick when they attend nursery or public school, where they are exposed to other children. Many parasitic diseases such as liver flukes and roundworms are spread through human feces and as population density rises, so does the incidence of such debilitating diseases, unless strict public health measures are enforced. Because of the artificially high local population densities made possible by our culture, we are sitting very high on a precarious pinnacle of sanitation and food-producing technology, quite possibly on the verge of some drastic density-dependent controls. Temporary failures of our complex technology, due to floods, hurricanes, earthquakes, war, or simple excessive demands because of overcrowding itself, bring immediate local threats of such diseases as typhoid and cholera.

We must not overlook the effectiveness of modern medicine. But as human population increases, the need for medicine to counter this potentially density-dependent factor will clearly increase. We also should not overlook another culturally inspired aggravation, due to worldwide transportation. Because of the ubiquity of airplane, railroad, boat, and automobile, a disease originating in some obscure corner of the world, where the local population may be somewhat immune, can quickly spread throughout the globe.* Such relatively minor Old World diseases as measles and chicken pox, carried by explorers and missionaries, devastated many of the nonresistant native people of the Pacific and the New World. More recently, "Hong Kong flu" and other illnesses have spread far beyond the regions that would have experienced them in earlier days. And whereas AIDS, for example, may conceivably be eliminated by prompt attention and Herculean efforts, it is also possible that it will spread throughout the world, whereas in previous eras it would probably have remained geographically isolated for generations.

There is also growing evidence that many animals respond to increases in their population density with changes in their behav-

*Syphilis was unknown in the Old World until 1495, when Columbus's crew—and quite possibly, the Great Explorer himself—docked in Genoa, Italy, and began assiduously infecting Europe with the fruits of their various adventures in the New World.

ior and physiology that ultimately have the effect of reducing their numbers. Once again, there is no reason to believe that such density-dependence represents foresight on the part of individuals or their species; rather, individuals (and individual genes) may maximize their personal reproductive success under conditions of crowding by behavior that, incidentally, tends to reduce the population. The relevance of all this to human beings is unclear, but challenging. John C. Calhoun experimented with rats, observing their responses to different population sizes. He placed five pregnant rats in a quarter-acre outdoor pen and provided sufficient food and water for the expanding population. During the ensuing two-year period, the five pregnant females could theoretically have produced 50,000 offspring, but the population never came close to that. In fact, it never exceeded 200, and eventually stabilized at about 150 animals. In a laboratory setup, with rats isolated from each other in small, self-contained cages, 50,000 animals and more could easily have been maintained in the same area. The big question then: how did these rats in a semiwild state keep their numbers so low?

The answer was found in their social behavior. They organized themselves into a dozen bands of about twelve or thirteen members each, with a single dominant adult male leading each band. Fighting between bands was common, and often disrupted normal care of the young. This in turn caused a high mortality among the pups and prevented the population from increasing. In fact, even the final population of 150 was an uncomfortably high number, maintained only by artificial feeding and the inability of the animals to escape and thus spread out. In nature, the population of an equivalent area would have been considerably smaller yet, not necessarily as a result of disease, predation, or cannibalism, but rather, simply because of the social behavior of healthy animals.

The rats in this experiment had apparently engineered a rather precise form of density-dependent population control that kept their numbers below the level at which more serious disruptions would be expected. But exactly what sort of disruptions? What would their behavior be like if they were constantly bumping into each other? To test this, Calhoun began his most famous experiment, in which most of the rats were eventually exposed to a population density about twice that which had been experienced by the penned animals in the previous test.

The results were striking. Under these high densities, the generally cohesive organization of normal rat social life broke down almost completely and with this collapse came heightened aggression, physical disabilities, and high mortality of adults (particularly females) and young. An epidemic of tail biting erupted, and younger animals were often seriously injured. Most important, normal sexual behavior and care of the young were greatly disrupted. Rats normally make hollowed-out nests in which the young are born. The crowded animals, by contrast, generally made inadequate nurseries, usually because they failed to collect the proper nesting materials. They were also sloppy housekeepers, often dropping their bedding and arranging it poorly. Because of this, the young frequently became scattered at birth and very few survived.

The crowded females were also very poor mothers, nursing inadequately and taking indifferent care of their offspring. For example, mother rats will normally retrieve their young if they are scattered about, usually transferring them to a new nest if the old one is disturbed. The crowded animals, by contrast, showed very little retrieval. They often dropped their young while transferring them, and frequently left them where they fell. These unfortunate pups were then commonly eaten by marauding bands of male rats. Such destructive cannibalism is never found among a troop of free-living animals, whose population invariably remains at a more manageable level.

Sexual behavior among rats normally entails a predictable sequence with mating success dependent on careful following of the correct routine. Males identify females by their odor, and are particularly sensitive to whether they are in heat. When the time is right, the male chases the female, who runs into her burrow, only to peek outside and watch the male, who responds by doing a little jig. When this is finished, the female reemerges and copulation finally occurs. Males in the overcrowded situation often failed to observe the basic amenities of rat courtship, following females into their burrows and chasing them in groups, behavior reminiscent of gang rape in human beings. With their sensitive reproductive mechanism fouled up, abortion and death among females was unusually high. This also retarded further population growth.

It is particularly interesting for students of human mental illness that the "sick" behavior that developed among these overcrowded animals was not homogeneously distributed throughout the population. Different individuals developed particular pathologies; four different categories of abnormalities were observed. The few dominant and moderately aggressive males appeared relatively normal. Others became "hyperactive," harrassing females and eating their young. Some became pansexual and mounted anything: males, females, young, old, receptive or unreceptive. One group was passive, taking no interest in sex or fighting, while the last group withdrew entirely and traveled about only while the others slept.

Calhoun's rats may be sending us a message: when social order breaks down, it seems likely that our worst biologic potential will express itself. Unnaturally dense populations can produce social abnormalities that ultimately bring down the population size. Assuming now that it is valid to extrapolate from rats to people—and this is something that must be done cautiously, although it is done routinely in biomedical research—then such density-dependent biological controls may be painting a grim future for a world of unrestrained human population growth, even if we are successful in avoiding catastrophic population collapse. If we are sufficiently wise, however, or sufficiently lucky, we will never know if our biology has endowed us with the capacity for similar responses.

We do not know the psychological basis for the various behaviors of Calhoun's rats. However, other studies have suggested a general mechanism that may have wide relevance. Biologist John Christian discovered that conditions of social stress, such as those brought on by local overpopulation, result in excessive demands on the adrenal glands. Sitting quietly above the kidneys, these small structures regulate a variety of essential chemical processes within the body, including mobilization of sugar reserves and resistance to infection during stress. If called into constant use, the adrenals grow. The demands placed on the body by their excessive functioning can result in shock and eventually adrenal exhaustion and death. The best example of this comes from a study that is now several decades old, but unsurpassed in its completeness and in its possible message for an increasingly overcrowded world:

In 1916 a few deer were introduced onto St. James Island, several hundred uninhabited acres in Chesapeake Bay. Their numbers soon increased to about 300, clearly an unnaturally high population. Fortuitously, Christian was already studying these deer when over half the animals died in 1958. The dead animals seemed paradoxically healthy except for one thing: their adrenal glands were greatly enlarged. By 1959, with the population at more normal levels, adrenal size was also back to normal.

It would be interesting to compare the adrenal size of human urbanites with that of rural residents. On the other hand, given the capacity of human beings to adapt, it is entirely possible that the resident of Chicago is no more stressed by his or her environment than the inhabitant of Maine's north woods; moreover, we seem to have a remarkable aptitude for stressing ourselves, no matter how or where we live.

If increased population size increases the stress on individuals, ultimately causing overwork of the adrenal glands and a higher death rate, this wholly biological mechanism could operate as a density-dependent check on population growth. The catalyst here would be personal stress, though, not population size as such. A population that is numerically dense could conceivably avoid stimulating the adrenal mechanism so long as the individuals concerned were not stressed by the presence of their fellows. Similarly, an objectively low population density could activate the adrenal stress mechanism if the individuals involved were somehow very sensitive to each other. It's all in the perception of crowding, not crowding itself. Christian discovered that adrenal size among woodchucks, for example, was greatest not when the population was highest, but rather when aggressiveness among individuals was most intense. This occurred early in the spring, at the mating season, and then again later in the summer, when the dispersing young were seeking to carve out living space at the expense of the adults.

We might anticipate that the symptoms of stress would be greatest among individuals at the bottom of the social hierarchy. After all, they are being "dumped on" by everybody else while unable to retaliate. In fact this has been shown repeatedly: there is a consistent correlation in many animal species between low social rank and excessive adrenal size. Again, possible correlations among humans have not been investigated, although a study of stress-related illnesses among employees of the Bell Telephone Company found that individuals lower in the company hierarchy, such as linesmen and operators, had a higher frequency of such illnesses than did members of the Board of Directors. We might have to revise our conventional wisdom regarding the "harried executive" and the happy-go-lucky workman.

Many animal species show another interesting phenomenon, relating social status to stress and ultimately in some cases to the regulation of population size. Low-ranking individuals are frequently denied access to females, either because they have been unable to establish adequate territories and are therefore not among the eligible reproductives, or because the females simply show a preference for high-ranking mates. This is especially true among animals such as grouse or turkeys that are polygynous (one male servicing many females). Under these conditions, subordinate males are truly personae non grata and may never copulate; in fact, they may actually be unable to mate even given the opportunity. Studies of subordinate males have revealed that among many species, testosterone levels are considerably reduced and testes are greatly shrunken by comparison with the dominant, actively copulating males. Ethologists refer to this phenomenon as "psychological castration."

Among the monkeys and apes, arrangements vary from species to species, making generalizations hazardous, but a clear analogy of psychological castration is not usually found. In fact, overt sexual rivalry among males is rare, and even among the despotic baboons, subordinates get to copulate. During the peak of estrus, however, when a female is most fertile and intercourse is most likely to produce offspring, copulation rights are largely monopolized by the dominant adult males.

Humans differ from most of the well-studied primates in living monogamously and ostensibly pairing for life. Since there are approximately equal numbers of men and women, our family system virtually assures that each male can have the opportunity not only to copulate, but actually to reproduce. There is therefore little chance for psychological castration among humans. However, the increasing problem of "psychological impotence" may be very similar in that it is often brought on by the stresses of modern life in which people are subordinated to goals and systems, if not to individuals as well.

Two British investigators, A. S. Parkes and H. M. Bruce, once noticed that when a recently pregnant female mouse was exposed to a strange male, her pregnancy was often interrupted by a premature abortion. It was subsequently discovered that the actual presence of the male was not necessary for this effect: a pregnant female will abort if she is placed in a cage where a strange male has been. This suggested that odor was responsible, a hypothesis that was later confirmed when females had their olfactory lobes surgically removed. Pregnancy in these nonsmelling animals was no longer interrupted by strange males.

It is not known to what extent such a pregnancy block mechanism operates among free-living animals, if indeed it does at all. However, it could indicate another density-dependent population factor, since as a population rises, the chances of pregnant females encountering strange males—or at least their chemical calling cards in urine or feces—would rise as well. For such individuals, it would probably be adaptive to refrain from a breeding attempt when the social system is disrupted and behavioral pathologies such as we have just described are likely. Given that most mammals live in well-defined social networks, the presence of a strange adult male could indicate that such disruption is under way.

Operation of the "Bruce effect" clearly requires a good sense of smell. Human beings, however, are unusual among mammals in possessing a poorly developed sense of smell. It is therefore very unlikely that odor-induced abortions would be activated by increases in human population. In fact, our relative insensitivity to smell has probably given us the potential of enduring denser populations more easily than most mammals. The stress generated by population density must be relayed to the animal by its senses. And smell, so well developed in most mammals, probably serves as a major avenue informing each individual of the immediate degree of crowding. Since trees are airy, windy places, long-distance olfactory communication has not been perfected among primates, and anatomically speaking, we are rather ordinary primates (discounting our enormous thinking cap, of course). Without a good sense of smell, we can probably experience moderate congestion with comparatively little stress. With it, we might well find auditoriums, elevators, buses, and probably our cities as well, utterly intolerable.

The various biological density-dependent controls to overpopulation such as cannibalism, aggression, automatic abortion, and stress cannot be acceptable to human beings. But for most animals and their component genes, they are ways of making the best out of a bad situation. That's why they have evolved. We could possibly derive comfort from the thought that such density-dependent factors might not operate among our species. But if we truly lack these controls, and also fail to exercise alternative cultural controls, we may be in even greater peril, since the various density-*independent* controls are in the long run no more pleasant or acceptable. Once again, we may be more endangered by the instincts we lack—or that we possess but can easily override—than by those few that we possess.

Lemmings often do something drastic when their population gets high. However, their famous "suicidal" rush to the sea is misunderstood. Lemmings do not actually commit suicide; rather, when the population density becomes excessive, they emigrate in large numbers, often swimming rivers if necessary. If instead of a river, lemmings reach the ocean, they treat this lethal barrier as though it is just another surmountable one. Perhaps if we were more subject to density-dependent effects, or if our culture had not progressed so far ahead of our biology, we would have a stable population today. As it is, we unfortunately resemble the lemmings, not in rushing to the sea, but in treating new conditions as though they are just like other ones, which we have always surmounted in the past. Suicide need not be intentional to be real.

Some animals do seem to practice an effective form of birth control. In the northern Arctic, snowy owls and the equally predatory, gull-like jaegers apparently adjust their clutch size to the amount of prey available. These animals feed on small mammals such as lemmings and mice, which as we have seen experience periodic fluctuations in numbers. The birds produce more offspring, sometimes even "double clutching" in times of plenty (a lemming outbreak), and fewer when food is more scarce. The small English birds known as "tits," relatives of our chickadees, feed on insects and require an abundant summertime prey population to obtain enough food for their ravenous young. Accordingly, they also adjust their family size to the amount of food available. Their eggs must be laid in the spring, however, when very few insects are out, in anticipation of conditions later in the summer, when the young have hatched and must be fed. How is this prognostication achieved? Insect populations later in the year are determined by temperature and moisture conditions during the preceding spring, and apparently these clever birds use spring weather as a cue, telling them how many eggs to lay. The schooling, once again, is by natural selection, which promotes the successful and fails the dunces.

These are appealing solutions to potential overpopulation, because they involve preventing overproduction rather than eliminating an excess. However, they are clearly automatic and biological, part of the animals' hereditary constitution. There is no reason to believe that we are similarly endowed. If we are to be similarly efficient, then we must do it via our culture.

We might consider a species to be a bathtub, with the water level indicating population size. The faucet is on and water is flowing in (births). But the drain is also open and some is going out (deaths). When the two are balanced, the water level remains constant; engineers refer to this type of system as a "steady state." Human culture has given us the partial ability to close off the drain: since we have not also turned down the faucet to compensate, the water is rising. Dangerously. Many species possess automatic faucet adjusters, or safety drains (density-dependent mechanisms) that keep the water from getting too high. If, as appears to be the case, we lack these devices, then the tub will certainly overflow unless we use cultural means to turn down the faucet. Paul Ehrlich has suggested that we can administer an intelligence test by seeing how someone responds when confronted with an overflowing bathtub: does he or she rush about with bricks and mortar, building up the sides, or, instead, turn off the faucet?

Increases in food production, medical care, pollution control, and improved housing are clearly helpful but are not permanent solutions. They are based on increase and growth, whereas ultimately we must attain equilibrium. Nothing else will do. We are adjusted for short-term success under conditions of resource abundance and high mortality, not for long-term equilibrium under conditions of resource shortage and low mortality. Moreover, because of our long evolutionary history as expansive creatures with an expanding population, we respond poorly to equilibrium as a conscious goal. Those who warn of the limits to growth seem dour, gloomy, and glum, as opposed to the optimists who cheerfully espouse "growth, growth and more growth," as Senator Paul Laxalt announced to the Republican National Convention in 1984.

Nowhere in the natural world is unrestrained growth a prescription for ultimate success, although it is a good description of cancer. Carried far enough, growth, growth, and more growth can lead only to death, death, and more death.

Under strictly biological conditions, different species can be imagined pushing as hard as they can against each other, and similarly for individuals within each species. With everyone pushing hard, anyone who lets up loses out. Moreover, no one suffers ill-effects for behaving as though unrestrained growth is the goal, since it can never be achieved. But thanks to cultural evolution, we have eliminated much of the natural resistance that our own potentially expanding numbers would otherwise have encountered. Our cultural evolution has been notably lopsided in this respect, since it has given us not only the science and technology to prevent early death, but also the religious and other cultural prescriptions to "be fruitful and multiply" as God is reputed to have enjoined Adam. Moreover, given our evolutionary history (both biological and cultural) as a species, we are especially susceptible to such advice. As a result, there is very little at the moment for us to push against, so that if we don't ease up—and wise up—we must eventually fall on our faces.

CHAPTER TEN

ENVIRONMENT

From Monkey Manure to Ecological Ethic

From now on it will no longer be enough to ask if man can do something. We must also ask whether he ought to.

—David Brower

As a description of our snowballing, deeply technologic, and not very thoughtful style of galloping cultural reliance, the word "harebrained" is particularly apt. If we don't destroy ourselves by war, our harebrained schemes may bring down the curtain on the human adventure simply by making our planet uninhabitable during peace, with an unhealthy assist from overpopulation. We are in danger of choking on the millions of tons of air pollution that darken our skies; certain rivers are so grossly contaminated they have been declared fire hazards; toxic waste dumps alone contain enough poisons to end life on earth. The litany of human-induced environmental hazards goes on: salinization, erosion, deforestation, desertification, acidification, artificial climatic changes from nuclear winter and ozone depletion to an overheated greenhouse effect with melting of the polar ice caps. We cannot feed the people now on the earth, yet the population grows ever larger. Wilderness—once an enemy, now a beleaguered friend—shrinks daily, never to be reborn. Endangered species slip away, forever.

And even if by some miracle, we are able to maintain a quantity of human life in the future, the quality of that life seems ineluctably diminished as we diminish the quality of the natural world. There is a special value in forests and mountains, running brooks, and quiet ponds, a butterfly in the spring and an elk bugling in the fall: these are the stuff of life, and that means human life, too. Let us therefore briefly examine the origins of our present environmental crisis, because it is a deeply human crisis as well.

Living things do not live in a vacuum. Neither do they evolve in one. The circumstances of existence for all animals and plants relate them profoundly to each other, so that the evolutionary advantage or disadvantage of any characteristic must be evaluated in terms of the organism *and* its entire environment. Thus, the speed of an arctic wolf is relevant particularly to the speed of its prey, the caribou, while the high-crowned tooth patterns of a horse are attuned to the abrasive nature of grasses, its preferred food. The intricate harmony of the natural world is due to elimination of misfits, combined with elegant elaboration of the successful forms and their interconnections. By most measures, Homo sapiens has been among the most successful of living things. Our population, after all, is high. Our mortality rates are very low. We occupy an extraordinary range of environments. We manipulate vast quantities of energy and materials. We offer an image of glowing evolutionary success. But present-day success is very different from future persistence, as the dinosaurs would tell us if they could.

Those dinosaurs, so often derided as failures, were great evolutionary successes in their time. And that time lasted more than a hundred million years, much longer than our own ascendancy so far. Moreover, despite their ultimate extinction, it seems that the dinosaurs were somehow "natural" in a way that we aren't. Most organisms, dinosaurs included, fit into their environment because those that didn't were promptly eliminated, as they still are today. Those that are left do not only belong, they are *part of* their environments.

Of course, this is not to say that species are immortal; they clearly aren't. Over long periods of time, extinction is the rule, as species are unable to keep up with changing environments. Something analogous can be seen in a fairly short timespan as well: consider, for example, the progressive changes occurring when a hardwood forest in the Northeast is cleared by lumbering or fire. The initial plant growth consists of grasses and various annual "weeds," followed by larger herbs and then shrubby vegetation. Finally after many years, an oak forest may appear, but since the large trees screen out much of the sunlight, the ground is dark, which prevents successful growth of young oaks to replace the old-timers. But maple seedlings are comparatively "shade tolerant" and are able to grow in the dark forest floor. They may eventually replace the oaks and a mature maple forest finally results. This final stage, ecologists refer to it as the "climax" vegetation, is self-perpetuating and will continue until human beings or some other disaster strikes once more. Then it all begins again.

Each of the progressive "successional" stages through which our forest passed possess characteristic plants and animals that eventually make conditions unsuitable for their continuance and so are replaced by the next stage. They literally carry the seeds of their own destruction. Of course, this elimination occurs on a local level only and is not true extinction. The various species involved generally survive, perhaps in another successional stage somewhere else. If the human species is creating an increasingly unlivable environment for itself, perhaps the same thing is happening to us, but with the notable exception that if the whole planet is rendered uninhabitable, we must all go extinct. Space fantasies aside, there is nowhere else to go.

We are unlikely to look on ourselves with equanimity as a passing stage in the progression of life, even though we have only been here a very short time, by the evolutionary calendar. Even aside from our very understandable emotional involvement in our own fate, something rings false about human extinction—and the human-caused extinction of other living things—as "natural." For one thing, there appears to be something peculiarly *un*natural about what we are doing to ourselves and to our world. We don't quite fit into the neat, interlocking system that governs other living things and their destinies. When we disrupt the lives and the futures of other living things, it therefore seems very different and much less acceptable than when they disrupt themselves. But why? In what sense are we "less biological" than a dinosaur or an oak?

The answer seems obvious: we are every bit as biological as any living thing, in that we were produced by natural selection and are subject to certain basic laws of the organic world. At the same time, however, we are also creators and creatures of cultural evolution. Once again, our glory is also our problem. And it is the crux of our present environmental crisis.

In a sense, we destroy our environment because we are able to. No other species of animal or plant poses a threat to the integrity of the earth, because none has the means to do so; any that made their environment uninhabitable for themselves are therefore no longer around today. As with weaponry, our destructive capacity does not derive from our simple biological characteristics. As animals we make only a small dent in natural communities, and, except for our extraordinary numbers, from the neck down we are interesting but unremarkable animals. But with the addition of tools, division of labor, language, and higher rational and technological capabilities, we have been *exploiting* nature in a way never seen before. Without cultural evolution, this would not be possible. With it, we have accomplished most of the things we prize as making us particularly human. And with it, like unruly Samsons, we threaten to bring down the earthly temple, crashing upon our heads.

Whenever animals and plants went extinct in the course of geological time, they did so because they could not adapt to a changing environment, not because they evolved themselves into extinction. Species suicide is unknown in nature, quite simply because the characteristics of living things are the result of biological evolution. Any genetically influenced tendencies that lead to decreased reproductive success would necessarily be selected against in the course of evolution, to be replaced by more benign traits. In the course of biological evolution, then, extinction occurs only as a result of happenings beyond the species' control.

There once were deer ("Irish elk") that possessed enormous antlers, exceeding ten feet in length. According to an earlier view, these animals went extinct because their antlers grew too large, getting hopelessly entangled in vegetation or perhaps so weighting their heads that the animals couldn't see where they were going and bumped into one another or walked over cliffs. This is very unlikely. If certain individuals had begun developing antlers that were disadvantageously large, they would have left fewer offspring than their more appropriately endowed brothers, and the average antler size would have decreased (a sort of negative feedback, evolution-style). More likely, environmental changes did in the Irish elk.

By contrast, human environmental abuse—whether indirectly as a result of population growth, or directly as a result of destructive and polluting technology—is not subject to the typical biological controls. It is mediated by our culture, not by our genes. Furthermore, even if the tendency for environmental destruction was under genetic control, as it would be in animals subject only to biological evolution, the pace of our culturally induced environmental destruction is much too fast for differential reproduction to have any hope of restoring an equilibrium. Cultural evolution has provided us, almost overnight, with the tools to destroy the earth, while once more (as in the case of aggression) denying us the kinds of inhibitions needed to restrict their use. Biology, acting alone, would never have granted us this awesome capability without also including protective controls; at least, not for long. But with cultural evolution acting unimpeded, it's a whole new ball game.

In general, biological evolution proceeds by relatively small steps, rather than large, discontinuous leaps.* The consequences

*Some biologists, such as Stephen Jay Gould, have emphasized that evolutionary steps may occasionally be larger than had been thought. But this is a matter of the degree of fine tuning. We are still talking of steps, not leaps.

of a particular trait, such as antler size, can therefore be evaluated by natural selection as it develops incrementally. By contrast, the enormous advances of human cultural evolution have provided us with great chunks of new characteristics, upon which evolution has had no opportunity to ruminate, and which in many cases must be either accepted or rejected as momentous technologic quantum leaps: the steam-powered locomotive, electronics, nuclear power, long-lasting pesticides, strip mining, acid rain. If some Paleolithic forebear had evolved a propensity to do something that was ultimately harmful to itself or its genes, natural selection would have nipped it in the bud. But if the trait was beneficial to itself, even if harmful to others, then it would have been favored. Moreover, if like nuclear weapons, it had the capacity of destroying the world, then it could hardly be selected against, since natural selection can only act on occurrences, not possibilities.

By a tragic irony of fate, we have once again been equipped with *fewer* biological inhibitions upon environmentally destructive behavior than have most animals. This may be because we are primates. As a group, primates generally do not make permanent living quarters. Those animals that do, such as most birds, are generally imbued with instinctive prohibitions against fouling their own nest. The young either back up to the edge of the nest and "let fly" over the side, or they produce special fecal pellets that are presented to the parent, who dutifully carries them away. By contrast, most monkeys and apes sleep in a different place each night. Like hikers unconcerned about leaving a dirty campsite because they will not be returning, our near-relatives show little reluctance to defecate in their own beds.

Ground-dwelling animals such as woodchucks or wolves are very careful to keep their living chambers clean and therefore free of possible disease. But urine and feces pass immediately out of a monkey's arboreal world and become someone else's problem. What, we worry?

Terrestrial animals can be located by predators who use their noses to trap the individual whose toilet manners were untidy. Not so for the primates. Monkeys thus make terrible pets, in part because they are notoriously difficult to housebreak. As recent immigrants to the terrestrial world, people—like monkeys—have few qualms about fouling their nest. It should cause no surprise, then, that for all our intelligence, it is harder to toilet train a human being than a dog.

Every new source from which man has increased his power on earth has been used to diminish the prospects of his successors. All of his progress has been made at the expense of damage to his environment which he cannot repair and which he could not foresee.

—C. D. DARLINGTON

MOST animals are not renowned for their foresight. But on the other hand, their capabilities of doing harm are also limited; their power is limited to their bodies, and hence their ability to damage their environment is quite limited. Predatory animals, for example, have to work hard for a living. They survive only at the expense of other lives and their victims can be expected to exert themselves to their utmost, in the hope of staying alive. Catching prey is thus not without risk; at the very least, it takes time and energy. Predators therefore tend to be conservative hunters, killing only what they need. By contrast, the fruit-, leaf-, and insect-eating primates evolved in the tropics, where different edible species are available at different times. Food is often "there for the taking," involving little risk or difficulty, and with little to discourage gluttony. In addition, primates do not store food against times of scarcity, presumably because they rarely experience substantial scarcities, and also because their preferred foods do not store well. It is therefore additionally likely that human beings have evolved with very few biological inhibitions when it comes to exploitation of natural resources, or planning ahead for a rainy day.

Biologist Garrett Hardin wrote a very influential scientific paper nearly twenty years ago, titled "The Tragedy of the Commons." In it, he pointed to the situation in Britain, when sheep-owners grazed their flocks on public lands known as the Commons. Although everyone suffered if the Commons was overgrazed, no one felt personal responsibility for maintaining it. Each shepherd also preferred to graze his sheep on the Commons rather than on his own, private land, and moreover, each shepherd recognized that if he refrained, out of concern for the common good, then someone else would simply overgraze. As a result, the would-be altruist found himself a victim of a type of Prisoner's Dilemma, forced to cheat (that is, graze on the Commons rather than cooperatively refrain), because if he didn't, he would be victimized in the long run by the selfish, overgrazing, noncooperating exploiters. So everyone felt constrained to go for all he could get, at the expense of the welfare of the environment generally. The tragedy of the Commons was not only the pressure that it exerted toward personal selfishness and egocentricity, but also the destruction of potentially productive land.

It is interesting to compare the human situation in this regard with that of two closely related members of the weasel family, both of which are superbly efficient exploiters of their natural environment, but neither of which possess the human capacity for environmental destruction or for conceptualizing their potential for triumph or tragedy. The otter is a wonderfully capable hunter of fish and invertebrates. Except where human activities have greatly reduced their prey populations, otters rarely go hungry. They can generally catch more food than they need, but they don't do so. Instead, they have become notoriously playful creatures, swimming after worried fish just to satisfy their frolicsome nature. But what they kill, they eat.

The mink, on the other hand, is also a masterful hunter but without the otter's playfulness. Mink find killing very easy, in fact hard to resist. They appear to violate evolution's logic by often slaughtering more than they need. However, their marsh environment is highly productive and in no danger of being overexploited. The mink's behavior has been unchanged for millennia and is fully integrated into the natural community. Crow, coyote, fox, and hawk thus take advantage of the situation, often consuming the mink's uneaten victims, gaining life by the mink's excesses.

Human beings are much more extravagant than the mink in our resource use. Our waste materials are generally used by organisms we consider nasty or otherwise disagreeable, such as flies, cockroaches, rats, mice, and the algae that pollute our lakes and rivers, feeding on the nutrients added as a result of human hyperactivity. These outbreaks are a far cry from the beautifully balanced system supported in part by the mink's behavior, largely because human sewage, garbage, and other such delights have not been around long enough for such an intricate biological web to have evolved.

Our penchant for excess is undeniable. "Man's chief difference from the brutes," wrote William James in *The Will to Believe*,

> lies in the exuberant excess of his subjective propensities—his pre-eminence over them simply and solely in the number and in the fantastic and unnecessary character of his wants physical, moral, aesthetic, and intellectual. Had his whole life not been a quest for the superfluous, he would never have established himself as inexpugnably as he has done in the necessary.

As James saw it, however, this was cause for celebration, not regret: "And from the consciousness of this he should draw the lesson that his wants are to be trusted; that even when their gratification seems furthest off, the uneasiness they occasion is still the best guide of his life, and will lead him to issues entirely beyond his present power of reckoning. Prune down his extravagance, sober him, and you undo him." It remains to be seen, however, whether we shall undo ourselves first, through the results of this extravagance.

Perhaps the most pressing environmental danger now facing our planet (nuclear war aside), is the increasingly rapid destruction of the world's tropical forests. These ecosystems are the most intricately balanced, species-rich, and poorly understood on earth. They are also irreplaceable, yet they are being destroyed at an extraordinary rate, partly as a result of lumbering but even more by clearing in order to raise beef, which in turn is sold to American consumers in fast-food restaurants such as McDonald's or Burger King. Thanks to the extravagance of human need and greed, Third World countries are permitting their precious tropical forests to be destroyed, while in fact, even the newly cleared grazing land does not retain its vitality for more than a few years before being degraded to cementlike "laterite," which is useless both for making Quarter-Pounders and for regenerating a lush forest.

The primitive slash-and-burn agriculture that preceded such large-scale deforestation was nearly harmless by contrast. During hundreds and sometimes thousands of years of such primitive environmental perturbations, the small areas that were cut down had ample opportunity to regrow. The large-scale environment was scarcely affected. But now, thanks to our penchant for excess combined with the technologic capacity to act excessively, we threaten permanent harm.

A small—or even a moderate—quantity of something may be innocuous, or even healthy. Excessive amounts, however, readily go beyond diminishing returns and often become troublesome. In small doses, such elements as zinc, cadmium, or nickel are necessary for life; in excess, they are poisons. Food is also necessary for life; in excess, it produces obesity. We would all die without oxygen; we would also die if we were forced to breathe 100 percent oxygen. Exercise promotes health; in excess, it produces injury. A smokestack may be a sign of jobs and economic vitality; a forest of smokestacks may be a sign of air pollution and an environment rendered toxic. Agriculture is a marvelous way of providing food for people; unrestrained agriculture—complete with deforestation, water table depletion, and widespread use of pesticides—could sterilize the planet. There is, in short, much to be said for the Platonic "golden mean." But human cultural evolution seems to lead eventually not to moderation but rather, to excess.

Basically, our environmental crisis has been fueled by three fundamental factors: First, the capability of producing environmental destruction (technology in a broad sense); second, the lack of inhibitions and integrated control regarding the exercise of this capability; and third, a particular attitude toward nature. Animals, of course, probably have no "attitude" toward nature, and it would make little difference if they did. Their mental abilities are too poorly developed for such abstract considerations and in any case, their lack of culture prevents them from acting effectively upon any "world view." By contrast, human beings typically do have an attitude toward nature and furthermore, they act on it, often very effectively.

This attitude is generally antagonistic and exploitative, at least in the Western world. Our capacity to exploit may even have diminished our need—and hence, our ability—to live peacefully with other Homo sapiens. Since we have the ability to "defeat" nonhuman nature even as we may be defeated by other human beings, then we may have taken the path of least resistance during much of our experience as a species. After a conflict, if the loser leaves and succeeds elsewhere, his and her descendants may eventually have inherited the earth, but part of that inheritance may include relatively little capacity or inclination for working things out with our fellow human beings. It is therefore at least possible that our success in subduing nonhuman nature, and thereby colonizing new areas, has empowered us to be (at least for the short term) the masters of our planet, while at the same time depriving us of the much needed ability to live more harmoniously with each other.

We see the earth as something to be conquered rather than to be appreciated, as challenging and threatening rather than nourishing and protecting. Wilderness is to be tamed, not savored. Wildlife is a "resource," rather than a legitimate fellow inhabitant of our shared planet. And once you have seen one redwood tree, you have seen them all. Caught in a fundamentally schizophrenic division between culture and biology, we also tend to dichotomize the world, seeing it in arbitrarily dualistic terms: subject vs. object, good guy vs. bad guy, man vs. nature. We live in two worlds, the biological and the cultural, and we therefore feel somehow separated from ourselves. We respond by feeling separated from nature as well.

"The ordinary city-dweller," writes philosopher Susanne Langer,

> knows nothing of the earth's productivity; he does not know the sunrise and rarely notices when the sun sets; ask him what phase the moon is in, or when the tide in the harbor is high, or even how high the average tide runs, and likely as not he cannot answer you. Seed time and harvest are nothing to him. If he has never witnessed an earthquake, a great flood or a hurricane, he probably does not feel the power of nature as a reality surrounding his life at all. His realities are the motors that run elevators, subway trains, and cars, the steady feed of water and gas through the mains and of electricity over the wires, the crates of food-stuff that arrive by night and are spread for his inspection before his day begins, the concrete and brick, bright steel and dingy woodwork that take the place of earth and waterside and sheltering roof for him . . . Nature, as man has always known it, he knows no more.

Although this isolation from our environment may be largely unavoidable, a natural result of our capacity for cultural development, to some extent such attitudes are also encouraged by our peculiar cultural systems. Thus, the Western world view is strongly colored by Judeo-Christian religious concepts, which are themselves dualistic. Strict alternatives are at the core of Western religion: God vs. His creation, sin vs. redemption, heaven vs. hell.

We are only rarely sensitive to the unity between organism and environment. Nor are we readily inclined to act so as to preserve the integrity of the system as a whole. Under the traditional view, we were given "dominion" over all the earth, with explicit instructions to multiply and subdue it. We face life with a chip on our shoulder.

Not only do we feel isolated from and antagonistic toward nature, we also "use" our environment to provide various complex gratifications: leisure, recreation, the experience of speed, of overblown material abundance, of control. An animal asks only for the necessities, in return for which it provides necessities to others. The primitive Stone Age human being and the enlightened Eastern mystic are alike in taking from nature only the basic necessities while refraining from grossly destructive pursuits. The latter does so because of his greater understanding, the former because he is unable to do anything different. The rest of us lean on cultural levers of great power to pry from nature our various personal and collective needs and wants.

As Max Weber pointed out, to the Easterner, rationalism means rational adjustment *to* nature; to most Westerners, by contrast, it means rational mastery *over* nature, a mastery that would be impossible without the levers of culture . . . and that may well be ultimately impossible in any event. Seeking to "win" in our self-proclaimed battle against our environment, to wrest satisfactions from a biological world that is equally intense—but often less successful—in seeking to retain its own structural integrity, we may yet defeat our environment. But if so, we shall lose ourselves.

Comforts and luxuries become "necessities" because they are accessible through technology and because our colleagues have them. In addition, the simple facts of economic survival often require environmentally destructive behavior; the modern world "makes a living" by mining the environment, sometimes literally. Doing so, we are committing an economic sin, if not a theological one: dipping into our capital rather than living off the interest.

In our violent treatment of the earth we often reflect our own aggressiveness. Ethologists identify something called "redirected behavior," as when an angry man pounds a table or slams a door because he is inhibited from attacking the real object of his aggression. We may similarly redirect much of our frustrated interpersonal aggressiveness toward our environment. It may be no coincidence that some of the world's most aggressive people, notably Americans and western Europeans, are also among the most environmentally destructive. In the process we are doing violence to each other, and ourselves.

Echoing David Brower's injunction, which appeared as an epigraph for this chapter, Nobel Prize–winning physicist Murray Gell-Mann pointed out:

> [I]t used to be true that most things that were technologically possible were done but that certainly in the future this cannot and must not be so. As our ability to do all kinds of things, and the scale of them, increase—for the scale is planetary for so many things today—we must try to realize a smaller and smaller fraction of all the things that we can do. Therefore an essential element of engineering from now on must be the element of choice.

Fortunately, there is some good news. Human beings, intelligent primates that we are, can exercise choice. We can overcome our primitive limitations and shortsightedness. We can learn all sorts of difficult things, once we become convinced that they are important, or unavoidable. We can even learn to do things that go against our nature. A primate that can be toilet trained could possibly even be planet trained someday.

CHAPTER ELEVEN

TECHNOLOGY

Plasm on Brass?

Oh, what a world of profit and delight,
Of power, of honor, of omnipotence . . .
All things that move between the quiet poles
Shall be at my command . . .
A sound magician is a mighty god.
—Christopher Marlowe, *Doctor Faustus*

To a weak-bodied, biological creature, limited to the productions of sinew and bone, any magician—sound or unsound—is a mighty god. And with our biology increasingly augmented by our culture, our abilities have become increasingly magical. It is another question, however, whether along with the power, we have gained honor or delight, never mind omnipotence. (Profit, at least for a few, is another story.)

The word "technology" derives from the Greek *techne*, meaning art or skill. It has come to refer especially to industrial arts, applied science, and practical engineering, although at least one definition, in the Random House Dictionary, is "the sum of the ways in which a social group provide themselves with the material objects of their civilization." Except for its emphasis on physical objects, this definition is not far removed from a definition of culture itself.

Technologies are assumed to be complex. In the late twentieth century, technology has come to be virtually synonymous with the latest in scientific applications, as in "high-tech" electronics (e.g., computers), chemical engineering (e.g., plastics), metallurgy, medicine, and/or nuclear energy. But we can speak equally of the technology of Paleolithic hand axes, basket-weaving technology among the Navaho, or the technology of propaganda.

Technology arises from the artful and organized application of tools. Tools, then, are the building blocks of technology, and are generally considered to be relatively simple instruments, usually hand-held, for performing mechanical operations. For a time, it was thought that human beings were distinguished from other animals by our use of tools, and the term *Homo faber* (Man, the maker) was even proposed as our scientific name. But as we have learned more and more about the lives of animals, this presumed region of human uniqueness has been progressively encroached upon, since we now realize that many different animal species employ tools. Some even make them.

The woodpecker finch of the Galapagos Islands selects a cactus spine, breaks it off to the right length, then uses it to probe for insects. (It doesn't have a long, thin beak like a woodpecker, so it has to manufacture an artificial one.) Egyptian vultures drop rocks on ostrich eggs, breaking their shells. Sea otters dive for abalones and also for a flat rock. While floating luxuriously on their backs, the ocean-going gourmets then position the rock on their chests and open the tasty abalones by pounding them against the mollusc-opening tool. Chimpanzees go fishing for termites with a long thin stick or blade of grass—the simple tool is inserted into a termite mound, termites grab hold of it, and the instrument is then withdrawn, the termites eaten, and the process repeated.

In some cases, such as termite consumption by chimpanzees, the tools are luxuries. In others—for example, the termites' reliance on the elaborate cooling, protecting, and humidifying properties of their complicated mound—the tools are absolutely essential. In the former cases (tools as luxuries among animals), we can easily imagine the creature without its tools. A chimpanzee is still very much a chimpanzee if it never goes fishing for termites just as a sea otter can still be a sea otter even if it dines exclusively on crabs instead of abalones. In the latter cases (tools as essential to the lives of animals), the tools are an indispensable part of the animals' existence, inseparable from other aspects of the creatures' biology. It is impossible to be a termite without a termite mound or comparable structure, although we do not consider that termites possess a mound-building technology.

Human beings, not surprisingly, are different. We have embedded ourselves in a heavy matrix of tools and technology, and yet, although we are utterly dependent on this aspect of our culture—no less than the termite—we and our technology are not inseparable. Whereas a termite is inconceivable without its mound, a person can readily be imagined without a lathe, a spinning jenny, a Linotype machine, or a home computer. In short, although we have produced elaborate technology on which we are increasingly dependent, our biology and our technology have remained largely independent of each other.

Traditional wisdom among anthropologists and biologists is that the use of tools was central to the evolution of modern Homo sapiens. There is a minority viewpoint, however, that holds otherwise. In *The Myth of the Machine*, Lewis Mumford argued forcefully that tools and primitive technology actually had a more limited role in human biological evolution than is customarily supposed. Mumford suggested that the crucial developments that made us uniquely human were "soft" and hence, not easily fossilized: language, religious belief, compassion and empathy, social organization, conscience and consciousness, moral and ethical systems. In short, it is at least possible that art preceded utility, and meaning supersedes mechanism. "The burial of the body," wrote Mumford, "tells us more about man's nature than would the tool that dug the grave."

The traditional focus on the formative role of tools and early technology in human evolution may well reflect a modern obsession with tools and technology, a felt need to justify and rationalize the Technological Person of the twentieth century. There is an old, corny song with the refrain: "You made me what I am today; I hope you're satisfied." If tools and technology literally made us what we are today, then perhaps we ought to be satisfied—with what we are, no less than with our present-day technology, since the latter is only the most recent incarnation of that which made us. But if, on the other hand, our fundamental essence is not the child of Pleistocene technology, then perhaps the proliferation of today's more recent technology is not necessarily the natural and appropriate outcome of our early apprenticeship as a species, but rather, perhaps, a mere aberration, a gratuitous and rather dangerous hyperextension of a capability which, although important, does not warrant the worship that it commonly receives. Bedazzled by our ability to build, maybe we have fallen victim to a new neurosis: the edifice complex.

More seriously, if technology often seems to make a monkey of us these days, it may be because in our hearts (or more important, in our brains and our genes) we are still monkeys.

A word is considered to be an example of onomatopoeia if—like "buzz" or "plop"—it was formed by imitating some external event that served as a direct referent. In an analogous sense the first tools can be considered as onomatopoetic. That is, they were probably conceived initially as simple extensions of the human body: the club a stylized and more powerful hand and fist; the bowl and pouch more efficient cupped hands; the flint scraper a heavy-duty fingernail; the knife a stronger, more maneuverable tooth.* If so, then special plaudits are due the first, unknown inventor of a tool that was truly independent of our own biological

*Psychoanalysts might also suggest that the spear is an aggressive penis substitute, and the basket a vagina. After all, the former is typically used by men, the latter, by women. But it is worth remembering that when Saint Sigmund himself was queried as to the significance of his fondness for cigars, he replied: "Sometimes a cigar is just a cigar."

equipment: perhaps the blowgun, or the bow and arrow.

Freed from their imitative and hence, limited role, tools led eventually to machines: from the Greek *machana,* for pulley (pulleys are one of the famous "six simple machines" and are made, in turn, from an arrangement of such tools as the wheel and a piece of rope). Dazzled as many of us now are by the discoveries of the Industrial Revolution as well as the high-tech of the twentieth century, it is easy to overlook the many and important early technologies which have loomed large in our civilized history, if not our evolutionary formation: painting, the potter's wheel, the loom, musical instruments, the plow, writing, kitchen utensils, the horse collar, watermills and windmills, and plumbing.

Lewis Mumford made a useful distinction between three levels of technology: eotechnic, paleotechnic, and neotechnic. Eotechnic (from the Greek *eos* or dawn) represents the first, primitive stirrings, the dawn of technology itself. The materials of eotechnic advances were stone, hide, wood, cloth, and simple metals. Paleotechnic advances were particularly associated with the Industrial Revolution; they involved the harnessing of enormous amounts of power, often with the combustion of fossil fuels: steam engines, iron smelters, and the traditional smokestack industries are all basically paleotechnic. Finally, neotechnology relies heavily on electricity, miniaturization, and efficiency; it is less obviously power-oriented and polluting.

The transition from eotechnology to paleotechnology to neotechnology has involved a progression of more work being done with less human labor. At the same time, it is interesting to note that the actual skill required may be less: it is easier to operate a machine that makes chairs than to carve them by hand, easier to tell time with a digital clock than with a clock with hands, easier to navigate by automatic pilot than by compass, and easier yet by compass than by the stars. As we learn new skills and techniques, we also lose old ones. Some of them may be no loss at all: if there are fewer and fewer people alive who remember how to start a car by cranking it, this hardly matters since cars are no longer built with cranks. And even if children no longer learn to tie their shoes, perhaps this too is no real loss, since now there is Velcro. But if cake mixes replace the skills of baking, and voice-activated computers make writing (or even typing) obsolete, there may be some difficulties if it is ever really necessary to cook at home, or if the electric power is interrupted.

In one of the most important American literary and sociological documents, *The Education of Henry Adams,* the scion of one of the nineteenth century's most illustrious political families developed the view that human history has been caused not so much by people acting on other people, as by forces acting on people. The force that according to Henry Adams personified the Modern Age was represented by a gleaming, powerful dynamo, displayed at the 1900 Chicago Exposition. Just as the Virgin symbolized the force acting on people during the Middle Ages, the dynamo represented to Adams the force of the modern, technologic age. "May the Force be with you," we were advised in *Star Wars.* To Adams, who lived during the transition from paleotechnic to neotechnic, the force is inseparable from us, and is not entirely benevolent, since it is responsible for a breakdown in personal relationships.

Gleaming gears, belching smokestacks, or sparking wires are not prerequisites for depersonalization. The highly organized slave societies of ancient Egypt and Babylon showed us that hundreds of thousands of laborers can be forcibly conscripted and highly organized, deprived of their individual rights and depersonalized in the extreme, with no more than eotechnology. Mumford identified such depersonalization as the most powerful and pernicious machine of all, which he named the "mega-machine." The myth of the mega-machine was twofold: that it was irresistible, possessed of Pharaonic deity as well as the momentum of history, and that it was ultimately benevolent. The myth of today's technology is no different.

⁂

> *Only let the human race recover the right over nature which belongs to it by divine bequest, and let power be given it; the exercise thereof will be governed by sound reason and true religion.*
>
> —FRANCIS BACON, *Novum Organum*

FROM Francis Bacon onward—and to some degree in fits and starts even before him—the Western expectation was that technology would solve our problems. Descartes pointed out that by the appropriate attitudes and actions, we shall "render ourselves the lords and possessors of nature," and speaking for the nineteenth-century Age of Progress as well as for his own, optimistic America, the poet Longfellow urged his fellow citizens: "Let us act, that each tomorrow finds us further than today."

The goal of progress, that place where tomorrow is to find us smiling, was sometimes stated to be the elimination of work and drudgery, as witnessed by Aristotle's expressed wish:

> If every instrument, at command, or from foreknowledge of its master's will, could accomplish its special work . . . if the shuttle thus would weave and the lyre play of itself, then neither would the chief workman want assistants nor the master slaves.

For others, the concept of technology was identified more diffusely with progress, which in turn, may be one of the most important and unappreciated legacies of Christianity. In contrast to the Eastern religions, paganism, or Greco-Roman philosophy, Christianity introduced the notion that the world was going somewhere, like an arrow shot from a bow rather than a static paralysis or an endlessly turning and returning wheel. The concept of progress is seductive and easy enough to grasp, perhaps in part because individuals progress through successive stages in their own biological, social, and intellectual development. Since Christianity teaches that human beings are born into sin, from which they must redeem their souls, progress seems all the more natural as well as desirable. Indeed, it is our human responsibility, the reason we were put on earth. "The education of the human race," wrote St. Augustine in *The City of God*, "represented by the people of God, has advanced, like that of an individual, through certain epochs . . . so that it might gradually rise from earthly to heavenly things, and from visible to invisible."

In short, the intellectual underpinnings were already in place when Francis Bacon, a thousand years later, heralded the Industrial Revolution and the Renaissance by extolling the goal of human existence in more secular terms: "the enlarging of the bounds of human empire to the effecting of all things possible." Progress, and progress alone (bound now inextricably, it seemed, to technology and the machine) became a new Western religion. It is likely, in fact, that the fervor with which progress/technology was embraced during the Renaissance and the following Age of Rationality was due at least partly to the collapse of the medieval belief in divine order and perfection. If the City of God was largely slums, and perhaps without even a city-planner, perhaps then science and technology could take its place. "The human race is continually advancing in civilization and culture as its natural purpose," according to Immanuel Kant, preeminent German rationalist of the eighteenth century, "so it is continually making progress for the better in relation to the moral end of its existence, and . . . this progress, although it may sometimes be interrupted, will never be entirely broken off." This new faith—faith in rationalism, science, and technologic progress—had replaced theologic faith as the prime draft-horse of Western civilization. But the underlying fact was unchanged: the hare, offering not so much eternal bliss in heaven as secular joys in the form of technology and progress, would still be humankind's savior.

There is no reason to conclude that such attitudes ended 200 years ago. Herman Kahn, the savant who gave us thinkable nuclear wars, also extolled "the curative possibilities inherent in technological and economic progress," prophesying a world of plenty if we would only embrace technology with yet more confidence and less restraint than we have shown so far.

Even some theologians have converted. Notable among them is the French priest/paleontologist Pierre Teilhard de Chardin, whose work has generated something of a cult following, especially since his publication of *The Phenomenon of Man.* Teilhard shows that you don't have to be a physicist or even a supply-sider businessman to love technology. He develops a humanist and theological argument that ends up celebrating technology as a manifestation of the power of collective human endeavor, such that the future human being will be part of a merging of the biological and the cultural, into something new on earth, the "Noosphere," a "domain of interwoven consciousness" in which all people will merge as a "single, hyper-conscious arch molecule." Urging us not to fight technology, but to join it so as to help bring about this final stage of human evolution, Teilhard rejects the "nightmares of brutalization and mechanization which are conjured up to terrify us and prevent our advance." He extolls the Manhattan Project for having showed that "nothing in the universe can resist the converging energies of a sufficient number of minds sufficiently grouped and organized," and excoriates those who "had the temerity to assert that the physicists, having brought their researches to a successful conclusion, should have suppressed and destroyed the dangerous fruits of their invention. As though it were not every man's duty to pursue the creative forces of knowledge and action to their uttermost end!"

Physicists such as Herman Kahn or Edward Teller recommend a growing technologic horizon simply because it is there, and because we can do it (moreover, if we don't surely the Russians will . . . and where would we be then?), with little concern for its impact on the human soul. By contrast, Teilhard urges us to achieve a mystical union and in the process, to expand our selves.

For others, however, the growing reach of technology offers not so much a promise as a threat. Simple tools and utensils—even simple machines—are, as we have seen, basically extensions of the human body. As such, they are unlikely to take on a life of their own. But in the progression from eotechnic to paleotechnic, and perhaps even more from paleotechnic to neotechnic, the hare confronts the tortoise with devices that are increasingly foreign and autonomous. Because of this growing independence, our creations seem increasingly to be a Frankenstein's monster, or a Sorcerer's Apprentice: forces outside our bodies, acting by their own power, and perhaps, even, by their own will. The computer is the apotheosis of such a transformation, notably HAL in the movie *2001: A Space Odyssey*, which is not only autonomous, but malevolent as well.

For those lacking the secular faith of Kahn or the religious faith of Teilhard, the creations of Homo sapiens are often less than entrancing. Although it is a product of our own actions, technology—by its fundamental strangeness from our biology—can readily feel as though it is imposed upon us rather than emanating from us. In any event, and as a result of consciousness as well as our technologic inventiveness, human beings are damned (or blessed) to live forever at the brink of Awe: for the primitive, awe at nature; for the modern-day technicist, awe at his and her own productions. During the dark days of the late 1930s, Jewish parents in Germany were often dismayed to see their children goose-stepping and "heil-Hitlering" with their playmates. Even in the concentration camps, some inmates became as brutal as the guards. Psychiatrists call it "identification with the aggressor," and the critic cannot help wondering whether our planetary embrace of technology does not also include some aspects of this, especially as technology increasingly takes on the mantle of a separate, autonomous entity . . . and moreover, a potential aggressor.

Unlike the moon, which only exposes its bright side to the earth, technology has another aspect. Let us now turn, therefore, to those who view its dark side.

Even now in the enthusiasm for new discoveries, reported public interviews with scientists tend to run increasingly toward a future

replete with more inventions, stores of energies, babies in bottles, deadlier weapons. Relatively few have spoken of values, ethics, art, religion—all those intangible aspects of life which set the tone of a civilization and determine, in the end, whether it will be cruel or humane; whether, in other words, the modern world, so far as its interior spiritual life is concerned, will be stainless steel like its exterior, or display the rich fabric of genuine human experience.

—LOREN EISELEY

TECHNOLOGY has done wonders to lighten the load of suffering humanity: for example, we can cure and often prevent diseases such as diphtheria, cholera, polio, typhoid, and whooping cough, whereas others, such as smallpox, have been eliminated altogether. So let's not undervalue the lowly technologist. Without the stonemason's craft and the glazier's skill, there would be no Chartres Cathedral, and without the instrument-maker, no Bach. Even in that extreme example of technology-run-wild, the nuclear arms race, technology has yielded some benefits as well: thanks to satellite surveillance, for instance, both the U.S. and the USSR can verify compliance with arms control treaties. The average twentieth-century human being, if fortunate enough to be at least moderately prosperous and to live in a technologically advanced nation, can be assured a reasonable chance of surviving infancy; growing up, he or she can contemplate the brilliant achievements of the world's past civilizations, and visit parts of the earth that were never available to our ancestors. We have more material abundance, with less work, than Homo sapiens has ever known before.

On the other hand, "goods" in the sense of material objects are not necessarily the same as "goods," the opposite of bads. (It is revealing that we use the same word for what should be two distinct concepts.) In widening a highway to adjust for excessive traffic, we often wind up stimulating more traffic so that instead of jammed two-lane roads we have jammed superhighways. In feeding an unprecedented number of people on our planet, the "green revolution" has spawned overpopulation that periodically leads to famines and ultimately, to environmental and human degradation. In producing the ultimate weapon, to keep the peace, we run a growing risk of losing everything, in the event of war. In short, there is a "flip side" to technology that isn't really very flip at all.

Although Philip II of Spain considered all inventors and innovators to be heretics, organized Western religion generally had less of a bone to pick with technology than with science. Thus, the Catholic Church resisted the heliocentric universe of Copernicus and Galileo, just as Protestantism opposed Charles Darwin; on the other hand, religion was more congenial to technology. In fact, it has been argued that the basic workplace virtues of thrift and punctuality were first developed in the very practical Benedictine monasteries of medieval Europe, from which they spread to the rest of the Western world. Those who dissented from technology's mandate—and its apparently overwhelming victory—were generally motivated by more secular concerns.

William Blake warned of the "dark Satanic mills" that accompanied the Industrial Revolution in Britain; paleotechnology in general, with its horrendous pollution, child labor, barbaric practices of near-slavery, and indifference to basic hygiene and fundamental human values, was not altogether beneficent in its impact on Homo sapiens, or on the rest of the natural world. As Lewis Mumford put it, while perfecting the mechanical art of multiplication, we neglected the moral and ethical art of division.

Even beyond the physical costs and dangers, and the social issue of justice and fairness, dissenters worried about the impact of technology on the human soul. Semiscientific visionaries, like H. G. Wells, foresaw a world in which the values of the machine depersonalized our species and ultimately destroyed it. *The Time Machine* painted a world in which romantic and humanistic values were embodied in the childlike and helpless Eloi, preyed upon by the cruel, rapacious Morlocks, machine-oriented troglodytes who ate human flesh. In their simplicity and passivity, the Eloi are reminiscent of de Tocqueville's warning, half a century earlier, that "a kind of virtuous materialism may ultimately be established in the world, which will not corrupt, but enervate the soul, and noiselessly unbend the springs of action." But in Wells's fantasy, the Morlocks no less than the Eloi were the victims of technology, condemned as they were to a brutal and cheerless underground existence as slaves to their machines.

Machines have grown in power and efficiency, becoming ubiquitous in the modern world. But although they have conferred many advantages, contrary to Aristotle's prediction they have not rendered slavery obsolete. In fact, the cotton gin actually increased the demand for slaves in the American South, and even with legal emancipation, the wage-slave is very much a reality today, consistent with Herbert Marcuse's observation that economic freedom should mean freedom *from* the economy. Moreover, the modern-day industrial worker, although not quite reduced to the nightmare of Wells's Morlocks, is increasingly becoming a latter-day pastoralist, a shepherd transformed into a machine-herd. "May not man himself become a sort of parasite upon the machine?" asks cyberneticist Norbert Wiener, "An affectionate machine-tickling aphid?"

The nineteenth-century liberal Russian intellectual and political troublemaker Alexander Herzen once predicted that technological advance in Russia would eventually produce "Genghis Khan with a telegraph." He was quite right: no better description has been written of Josef Stalin. But now, things are infinitely worse: Genghis Khan has nuclear weapons. Worse yet, Genghis Khan is transnational, found in Washington no less than in the Kremlin, in Belfast no less than in Beirut.

We have unleashed not only the atom, but the aggressive, inventive, and insatiable curiosity of Homo sapiens upon a world that previously knew only the feeble ploddings and proddings of strictly biological creatures.

Our reach, according to Robert Browning, should exceed our grasp, "or what's a heaven for?"—the poet enjoins us to strive to accomplish more than we are capable of achieving. Ironically, the situation has now reversed itself: our grasp has exceeded our reach. We have a technologic tiger by the tail, right here on earth. As for the Sorcerer's Apprentice, things are getting out of hand; or, like Genghis Khan with a telegraph, a Gulag, napalm, nuclear weapons, or a bulldozer, the artificially extended reach of our hands has exceeded our ability to achieve the necessary hand-mind coordination. We may well have accomplished more than we are capable of mastering. Like Halvard Solness, the tragic master builder in Ibsen's play of the same name, we have built higher than we can climb.

Sometimes, it all threatens to come tumbling down, and not just in war or intentional cruelty alone. Instead of "energy too cheap to meter," we have nuclear power plants that are economic basket cases, generating radioactive waste that will remain fiendishly toxic for many times longer than human civilization has yet existed. Instead of making the world a "global village," worldwide communications now make potentially lethal misunderstandings instantaneous; being able to talk with other nations has not given us anything more to say. The modern airplane, giving us convenient travel, has given us the bomber as well. The ability to transport food quickly and over long distances has resulted in the centralization of bakeries, for example. So whereas the average Frenchman can buy fresh-baked *petit pain* at a *boulangerie* within walking distance of his or her home, the average American has the privilege of buying prepackaged white bread, baked perhaps hundreds of miles away, and laced with preservatives to help the product survive its transportation from point of manufacture to wholesale warehouse, to supermarket, to breadbox. And when it comes to the daily grind of getting to work, although it is undeniable that our high-speed people-moving technology moves more people, more quickly than ever before, it is uncertain whether in fact the average commuter actually spends less time commuting than his or her counterpart several hundred years ago; having given ourselves the ability to travel long distances quickly, we also choose to live proportionately farther from our workplace.

At other times, it really does come tumbling down. The industrial disaster at Bhopal, India, in December 1984 is a classic, tragic example. At least 2500 people were killed, perhaps as many as 10,000. A few weeks earlier, liquid natural gas had exploded in Mexico City, with about 500 fatalities. Sixty thousand people had to be evacuated from the vicinity of the Three Mile Island nuclear power plant in Pennsylvania, in the aftermath of a leak and potential meltdown during 1979. Toxic wastes have ruined lives and threatened many more in Seveso, Italy, and Love Canal (near Niagara Falls, New York). In 1971, cargoes of Mexican wheat and United States barley, distributed in Basra, Iraq, had been treated with a mercury compound to prevent spoiling. The wheat was intended for use as seed, but the recipients, misunderstanding its purpose, consumed large quantities of the poisoned grain: as many as 6000 may have died. And there are asbestos, PCBs, Agent Orange, oil spills, and sewage leaks, to mention just a few. The disastrous explosion of the space shuttle Challenger was a particularly spectacular example of technology gone awry. It was troublesome for many Americans, not only because they personalized the catastrophe, but also because it represented the failure of a high-tech program that Americans had begun to take for granted.

In each case, the usual response is to blame operator error, mechanical design, faulty communications, insufficient local oversight, and so forth—certainly not the conflict between cultural and biological evolution, or between human capabilities and human need on the one hand, and technology and human greed on the other. According to political scientist Robert Engler, one historian estimated reassuringly that of those killed at Bhopal, one half "would not have been alive today if it were not for that plant and the modern health standards made possible by wide use of pesticides" of the sort manufactured there. The gods of technology giveth, and they taketh away.

Just as psychoanalysis has been described as treatment for the worried rich, so has concern for the environment and anxiety about technology. On this view, interestingly, the far right and the far left have sometimes converged, the former eager to justify maximum profits and to defuse public criticism and scrutiny of corporate irresponsibility, and the latter eager to enhance employment in today's smokestack industries. It is therefore noteworthy that the real victims of technology run wild, those most likely to be taken by this unleashed devil, are in fact the hindmost of society.

The severest industrial tragedies have not occurred in the United States, but in Third World countries, which often have environmental and worker protection laws that are notably lax compared to those of America, in an understandable effort to lure foreign investment and jobs. These industrial plants are typically located in slums, where land and labor are cheap. And where the effluent doesn't bother the affluent.

But eventually, the chickens must come home to roost, for the wealthy nations as well as the wealthy within each nation. "Of 600 registered generic pesticides in common use today," writes Jonathan Lash in *A Season of Spoils,* "79 to 84 percent have not been adequately tested for their potential to cause cancer, 60 to 70 percent have not been tested for their potential to cause birth defects, and 90 percent have not been tested for genetic mutations." The public has a legitimate, gut fear of catastrophe, and will hold its collective breath in fascination over towering infernos, nuclear meltdowns, and Titanic disasters, whether real or celluloid. By contrast, slow-acting substances, which accumulate over time and may exert their toll only over a span of years, simply don't register very high on humankind's Richter scale of ground-shaking events.

There is a real ambivalence here, not just an ambivalence of attitude, but one of substance as well. We are like the victims of a bad marriage, who nonetheless love each other: we can't live without technology, but we are having more and more trouble living with it as well. (Similarly, we can't live without our biological self, but find it difficult living with it as well.) Contemplate a sleek Lamborghini sports car, a finely tuned Jaguar XKE, or a smoothly humming Cadillac El Dorado, all triumphs of high technology—complete with electronic ignition, integrated circuits, computer-assisted carburetion, high-impact plastics, and space-age alloys. And then consider that they each rely on squashed dinosaur guts (that is, petroleum) to keep them going.

Technology is culture incarnate and yet it relies fundamentally on biology, no less than we—ourselves the incarnation of cultural evolution—rely on our own creaturely strengths and weaknesses. More than fifty years ago, W. F. Ogburn wrote *Living with Machines,* proposing the concept of "cultural lag." According to Ogburn, human subjective values (art, religion, ethics, emotion, even politics) have failed to keep pace with the innovations of machines. As Ogburn saw it, our human responsibility was to reduce that lag by speeding up our adjustment.* More recently, theologian Harvey Cox wrote that "the secular city signifies that point where man takes responsibility for directing the tumultuous energies of his time." Since technology has rendered most social and political structures obsolete, it is our job, as inhabitants of the "Secular City" rather than the City of God, to weave a "political harness to steer and control our technical centaurs."

But the problem is not just technology alone, or the tendency of human culture to lag behind a technology that is growing increasingly inhuman. Rather, it is the tendency of culture to outrun biology. Moreover, as we have seen, technology is a two-edged sword, offering benefits as well as extracting costs. We clearly need more than a dyspeptic rejection of modernity, we need instead an insistence that technology be measured by how much it contributes to life, rather than evaluating life by how much it contributes to the furthering of technology. Technology itself is not the enemy. The difficulty lies precisely with the creators and wielders of

*In "Sociology and the Atom," Ogburn gave an example of what he meant. He proposed that society's response to the challenge of the Nuclear Age—our way of diminishing the cultural lag that had been created—should be to dissolve our urban civilization and reconstitute the United States as a semirural society based on thousands of small towns dispersed around the countryside. It was a preview of the "crisis relocation" plans of the early 1980s.

technology; that is, with Homo sapiens, that confused and confusing creature of whom we are so justly proud and about whom we are so legitimately worried.

Earlier, we briefly considered the relationship of technological evolution to social evolution, pointing out that the latter tends to be much more sluggish, undirectional, sometimes circular, and perhaps as often regressing as advancing. As the "centaur" that has provided much of the brute force for human societies, technology has not been notably controlled or intelligent, but rather, something of a muscle-bound idiot. There is a more hopeful vision, however, since technology does progress, and—at least in theory—it can be harnessed and directed as we, the drivers, see fit. We cannot realistically accelerate biological evolution, nor can we count on altering nontechnological social evolution by contemplative philosophy, or fervent appeals to morality. But we can alter and orient technology; indeed, we do so every day. Population control can be achieved by appropriate uses of technology, and it seems likely that even the human desire for large numbers of children can be altered by technological advances that render child mortality reliably low. Polluting technology can be replaced by nonpolluting alternatives, and we can even engineer microorganisms with the genetic capability of degrading oil spills, plastics, and noxious chemicals, and particle accelerators to deplete plutonium and uranium-235. More efficient houses, industries, and automobiles, based on recycling and renewable resources, can be designed; to some extent, this is already being done, although on a small scale.

No matter how much our biology and even some of our social structures may resist, it is at least possible that we can use technology artfully and intelligently to orient our lives in healthy ways. If so, then today's manifest conflicts between technology, nontechnological culture, and biology may be transient and in the long run surmountable. Perhaps it is not too late for a reconciliation, with advances in science and technology serving as one possible fulcrum around which such harmony could be established.

Technology alone won't save us, any more than technology alone will destroy us. The missing piece in the puzzle is the human being: biologically evolved yet culturally and socially conditioned. And that cultural and social conditioning, by definition, is something over which we have power.

Many people have written about the "technological imperative," the notion that if something can be done, it must be. The implied image is of a large genie, arms akimbo, scowling and tapping his foot impatiently while waiting for us to do what eventually we must. Perhaps, however, we might occasionally show that genie who is the boss.

We have often assumed a confrontational attitude toward nature, frequently to a fault. We climb mountains because they are there, but we certainly haven't accepted the legitimacy of something—whether social injustice or smallpox—for the same reason. Why, then, should we accept the dominance of technology and machinery, just because they are there? The irony is all the more pronounced when we consider that unlike nature, technology exists only because we have created it. Rather than accept technology as an immutable given, and seek to rectify cultural lag by establishing human institutions that are more congenial to technology, perhaps we might consider establishing technologies that are more congenial to human beings, and to the rest of the planet.

"The Walker," in Robert Frost's poem, comes upon a nest of turtle eggs on a railroad track. He muses that

> The next machine that has the power to pass,
> Will get this plasm on its polished brass.

The polished brass of a railroad engine is a universe away from soft and squishy turtle-plasm. It is built of sterner stuff and is indifferent to the latter's fate. Only because of the intervention of another kind of squishy stuff—human-plasm—are the two coming into tragic contact; without that human-plasm, the polished brass wouldn't exist at all. So the world of polished brass and the world of plasm may not really be that far apart after all. We can look the other way when the train hurtles past, or shake our fists in impotent rage, but perhaps our responsibility lies deeper, and is more demanding: to climb aboard, and be a prudent engineer.

CHAPTER TWELVE

ALIENATION

Homo extraneus

I, a stranger and afraid
in a world I never made.
—A. E. Housman

Homo sapiens could also be called the alienated animal, estranged from much of his world, and from himself. We are probably unique in nature, the only animal that is somehow out of place in its own surroundings. Twentieth-century literature, poetry, theater, and cinema all reflect a rising tide of alienation. But whereas the strangeness of human beings in their own environment has probably been on the rise in recent years, it has actually been with us for a very long time, once again a result of our uncoordinated culture and biology. We are animals, living in a man-made environment. Small wonder we *feel* strange: we are strange.

There are many symptoms. Mental illness, anomie, frustration, boredom, antagonism, withdrawal, insensitivity: all these and more derive at least in part from a disharmony with the world we inhabit. (Including in this world, of course, our fellow human beings.) The human environment is a product of our culture, and within this complex edifice, a biological creature must live. Insofar as we possess any genetic basis for our behavioral tendencies and inclinations, this biological foundation must be largely appropriate to a pretechnological if not precultural environment. In effect, we entered a time machine just after we discovered tools, fire, agriculture, technology, and the rest of its offspring . . . and the dizzying pace of our trip has rendered us increasingly disoriented ever since.

"The heart of our difficulty," to historian Arnold Toynbee, "is the difference in pace between hare-swift movement of the scientific intellect, which can revolutionize our technology within the span of a single life-time and the tortoise-slow movement of the subconscious." In sketching the origins of alienation in the conflict between the hare and the tortoise, we shall consider first the discordance between human beings and the environments that our culture has enabled us to produce. Then, we shall look briefly at the alienation engendered by consciousness itself.

In a sense, virtually any human environment, even the most pastoral, must strictly be considered artificial in that it shows the unmistakable impress of ourselves and our culture. But perhaps no human environment compares with our cities when it comes to blatant disregard for certain aspects of our biology. Dependent upon the countryside for food, raw materials, and amelioration of their pollution, cities nevertheless have a life of their own. They are wholly manmade, where people are often crowded beyond belief and without relief, and in which one could easily spend a lifetime without walking on the earth or sitting under a tree. Needless to say, this is a peculiar setting for a living, breathing, sweating, biologically evolved creature.

It is also a very recent development: by 1800, only 50 cities in the world had populations of 100,000 or more. By 1985, there are more than 1500, of which more than 150 have one million or more residents. Mexico City alone is predicted to have 30 million inhabitants by the year 2000. To Plato, the size of a city was limited by the number of people who could hear the voice of a single orator; thanks to electronics and telecommunications, that number is now infinite. But it is another question whether our capacity to tolerate such numbers and such densities matches our ability to communicate and accumulate.

Although many people seem well adjusted to cities and would be loathe to leave, the truth is that our cities have enormous problems, most of them resulting from the disparity between our cultural creations and our biological needs. The problems of aggression and social disorganization were discussed earlier; while generally applicable to the human situation, they are also especially appropriate to humans in cities. There are, in addition, other alienating factors that are more specific to cities.

We all know the popular stereotype of the country bumpkin who is hypnotized and bedazzled by all the hustle, bustle, and bright lights of the Big Town. But we are all country hicks at heart. Cities present situations of excessive social stimuli—sights, smells, sounds—all ever-present, ever-changing, ever-demanding, urgently assaulting the senses. And there is little escape. One recourse of the city dweller is a retreat into what the religious philosopher Martin Buber called an "I-it" attitude toward our surroundings and our fellows. The alternative, "I-Thou," is a deeper, loving relationship in which the meaning of one partner is predicated upon the other, both achieving a transcendence of their otherwise limited selves. By contrast, the I-it relationship implies a thoroughly objectified attitude in which the individual sees him or herself as completely encapsulated within his or her skin, always distant from the other, from "it."

The city's characteristic bigness, noisiness, constant flux, and impersonality make I-Thou relationships unlikely. The average urbanite will encounter literally hundreds or thousands of people every day, and almost every one of them will be a stranger. This is a commonplace observation, but in a sense, a highly significant one. We meet strangers every day, not only passing them on the street, but crushing against them in public conveyances and doing business with them all the time. Visit a primitive human culture today in one of the backwaters of the world where homogenizing

Western culture has not yet penetrated: the people may be frightened, shy, antagonistic, or intensely interested, but they certainly won't ignore you.

You can be sure that the average tribesman in the mountains of New Guinea or the deserts of the Kalahari does not often meet with strangers. Regular dealings are almost entirely with relatives, friends, or at least acquaintances, and the chances are good that it was the same for our ancestors. The contrast with the average city dweller is extreme, and so are the results. Knowing a person is different from knowing a fact, a distinction that is oddly absent in English but well expressed in French, with the different verbs *connaître* (to know someone) and *savoir* (to know something). The former takes longer, is more difficult and more meaningful. When people know each other, they act differently than when they first meet. Formalities decline, protective mechanisms are relaxed. There is a mild but definite strain that comes with meeting a stranger. His or her attitudes and reactions are unknown. Although the circumstances of meeting generally indicate nonhostile intent, and often provide much additional information about what to expect, a certain biologically appropriate suspicion may well generate initial malaise. And this must become a constant part of the city dweller's life.

Even when introduced to someone at an intimate social gathering where personal interaction is expected, the overwhelming majority of people have great difficulty remembering their names, typically because they are stressed and preoccupied seeking to respond to the stranger(s).

One way to lessen this strain is to keep others outside the protective envelope of our existence. Actually, we cannot possibly know everyone in our path when we walk down a busy urban street. We cannot personally greet everyone on a crowded bus. We cannot reach out to the humanity that flourishes around us, breaking upon us like the ocean upon a rock. We cannot respond in the deeply human way for which biological evolution has undoubtedly prepared us. Rather, we must keep the world and each other at arm's length; although Buber saw it somewhat differently, I-itness is a defense mechanism, born of sheer necessity and helping us keep our emotional balance in a world made unstable and chaotic. So we surround ourselves with an impermeable barrier of indifference, sidestepping our fellows in silence, ever careful to avoid recognition of the other as a human being. In the New York City subways, no one looks anyone in the eye. Out on the street, we may even step over a fallen body, or observe an attempted murder with a cold indifference and reluctance to "get involved."

"Men, it has been well said, think in herds," we learn from a nineteenth-century treatise, quaintly titled *Extraordinary Popular Delusions and the Madness of Crowds.* "They go mad in herds. While they only recover their senses slowly and one by one."

In a sense, the urban dweller cannot be faulted for such behavior. It is forced upon him and her by the insane nature of the crowded madmade environment. Cities differ with respect to their encouragement of this attitude: New York City, for example, is one of the worst. Because it is almost impossible to find a comfortable place to sit, a drinking fountain, or bathroom facilities, its inhabitants are forced to use the city as a place between goals, a means to more private ends. They rush about, stone-faced and seemingly always in a hurry, callous to whatever or whomever they may encounter.

Many animals indulge in greeting and so-called "contact" signals, which also serve to reassure subordinates and appease the dominants. A flock of sparrows, like a crowd of people, is a noisy affair. Each bird periodically makes a short vocal call which informs the others of its presence and helps maintain optimum spacing among individuals. It also probably serves to reduce aggression between neighbors, informing them that all are bona fide members of the group. Among ground squirrels and prairie dogs, strangers sniff each other's faces in a stereotyped greeting ceremony. Dolphins chatter to each other almost constantly and chimpanzees extend their hands. Humans also shake hands when the meeting is formal or direct; when simply passing an acquaintance we generally utter a brief "hi" or equivalent, and/or smile and nod our head. These greeting ceremonies may appear inane or useless, but a brief experiment in withholding such signals will quickly reveal their very necessary function. We arouse quick suspicion and antagonism if we fail to greet our friends and acquaintances. Why? And what relevance does all this have to alienation and urban life?

The meeting of two animals produces immediate tension, especially if the encounter is unexpected. Most animals have therefore evolved means of reducing this tension, thereby preventing undue disruption of normal behaviors. It is also desirable to indicate association with others when such exists, so friends greet each other. Although the exact nature of these greetings varies among people depending on the specific culture, the general behavior is universal. Walk down a street of any small town: you will greet or be greeted by virtually everyone you meet. This is because with a small population, everyone knows everyone. Now walk along a busy city street: everyone studiously refrains from engaging anyone else. To do so would be physically impossible as well as physically dangerous. We can only speculate on the unresolved stresses generated by such unfulfilled encounters as we withdraw into ourselves. Protecting ourselves in one way, we stress ourselves in another.

One obvious result of the city's enforced anonymity is its crime rate. Unlike impulsive violence, in which the victim and perpetrator are often members of the same family or at least well known to each other, premeditated robbery or burglary is almost always directed at a stranger. A small-town resident doesn't rob the corner grocer; everyone knows nice old Mr. McPherson. But if McPherson is a nameless, familyless, disembodied, and anonymous spirit in a big city, he can be attacked with relative ease. Beyond this, the trend toward major corporate or conglomerate ownership makes the individual units all the more vulnerable. No one knows Mr. Safeway, but McPherson is a person. As our culture forces us together in such unnatural densities and proximities, the same defense mechanisms that defend us by maintaining emotional distance and indifference deny our natural inhibitions the opportunity to protect us.

It appears that as population density increases, the value attached to each individual decreases proportionately. We move from what in German is known as a *Gemeinschaft,* a society based on face-to-face personal ties, to a *Gesellschaft,* one characterized by impersonal, commercial relations.

The specter of galloping Gesellschaft provides a cogent argument for population control and for reducing urban concentrations, or at least encouraging the development of local neighborhoods where a healthy sense of community and of personal worth can still survive. It is ironic, for example, that one way for people to become suddenly very important is to go into the wilderness, away from people. On a backpacking trip where companions are rare, a stranger encountered on a lonely trail can be a delightful and valuable experience. There is often a lot to discuss: destinations and starting points, trail conditions, weather prospects, the location of good camping and watering sites, wildlife observed, etc. The same people, meeting at Fifth Avenue and Forty-second Street, would hardly respond to each other's presence. A good reason for getting away from the crush of people, if only temporarily, is to counter the city-bred alienation of people from each other.

Given all the disadvantages of city life, why do so many choose it? In most cases, there seems to be less a conscious selection of the city per se than a following of jobs, convenience, and economics, all of which may lead to the city. There is also, of course, the simple accident of those born to parents who were themselves previously lured by one of these circumstances. But other positive inducements exist as well. Earlier, we discussed the ethologists' concept of releasers and their potent artificial counterpart, supernormal releasers. These are manmade stimuli that elicit an exceptionally strong response by exaggerating certain characteristics of the normally occurring releaser. The city may well constitute a supernormal releaser for human beings, a cultural hyperextension of our fundamental sociality.

Like most other terrestrial primates, we are gregarious creatures, although our primitive social units were undoubtedly much smaller than the modern metropolis. We banded together for hunting, mating, child rearing, food gathering, defense, the transmission of culture, affection, and companionship. Evolution must have selected strongly against solitary tendencies, as it still does today among most primates, while favoring the social creature who was attracted to communal gatherings. The asocial person, like the solitary baboon, did not survive long and left few descendants. People would doubtless gather at a safe camping place, much as baboons do today at protected sleeping trees, or around the spoils of a successful hunt. For some animals, such as fish in a school or possibly primates in a troop, there may have been safety in numbers for purely statistical reasons as well. If a predator is likely to take, say, the first three individuals that he or she encounters, there would be a benefit to huddling closely together, not so much out of fondness, but in hope that as a result, someone else will be chosen rather than yourself. Not surprisingly, then, a school of fish becomes more tightly bunched when a barracuda swims near, and a flock of ducks flies even closer together when a peregrine falcon swoops overhead. British biologist W. D. Hamilton, who elucidated the mathematics of grouping for safety, called his scientific paper "Geometry for the Selfish Herd."

It may be an overstatement to describe a human city as a selfish herd, and certainly, most urban dwellers don't flock together out of conscious fear that a leopard might leap out of a dark alley, especially eager to make off with an easy, solitary target. But there is no doubt that we find safety, and reassurance in numbers. And if groupies were more fit than loners, the various signs of a group may well have been made attractive to us by biological evolution. However, within the last few thousand years, cultural evolution, impelled by the discovery of agriculture and following the path of least economic resistance, has spawned enormous gatherings of people, greatly surpassing anything we had "naturally" produced. The color, the noise, the variety, and the sheer excitement are fascinating and almost irresistible. Like moths to a flame, we have been drawn to this oversized stimulus, this supernormal releaser.

John C. Calhoun's experiments with overcrowded rats also suggest a far-reaching interpretation of human city-seeking. His rats, showing a wide gamut of hideous responses to excessive population density, were not actually forced to live so closely together. They *chose* it. The experiment was designed so that they had to feed from central bins from which each animal could only obtain a small amount of food at a time. Consequently, they spent a long time feeding and since there was room for many rats to feed together, they soon became accustomed to feeding alongside each other. Soon they became "conditioned" to each other's presence, actively seeking the larger group, even though the resulting overcrowding produced what are (at least from our viewpoint) very unpleasant results. Calhoun labeled the end product a "behavioral sink"—the unhealthy connotations were intentional—and he referred to his animals' intensive sociality as "pathological togetherness."

Yet there is hope. Eric Hoffer emphasized that many of our most valuable social and cultural advances came from cities. And as shown especially by Anne Whiston Spirn in her recent book *The Granite Garden,* many urban problems are in fact soluble. The Eternal City is doubtless a self-serving myth; the infernal city, however, need never become a reality, if we recognize cities as real environments, with the potential of being planned, humanized (sometimes even naturalized), and made tolerable and at times, even delightful. In our haste to produce supernormal releasers, it may be that we have too quickly given up on cities as environments. Spirn details the benefits that can result once we take account of the water dynamics, plant and animal life, soil compositions, wind patterns, and heat vectors that operate within so special an environment. Even ancient Rome met its citizens' needs for water, and modern-day Stuttgart has done a remarkable job adjusting human-made industry to nature-made airflow so as to minimize air pollution. After all, times have changed: once, we took our environments as we found them, or went elsewhere. Now, we have the capacity—indeed, the obligation—to make our own environments. When we begin doing this seriously, and not simply with an eye to economics, we may find ourselves feeling less like strangers in a strange, artificial land.

Aside from the alienation induced by our living places, notably our cities, there are the problems arising from the fact that we have surrounded ourselves with the increasingly strange products of our own creativity. When scientists wish to deal with devices they do not understand, they call them "black boxes." We know what goes into a black box and what comes out of it (inputs and outputs, to the engineer), but we don't know what goes on inside. Most psychologists, for example, treat the brain as a black box, with stimuli going in and behavior coming out. With the advent of increasingly elaborate technology, human beings are finding themselves increasingly surrounded by black boxes. We wake up in the morning, flick a switch (input), and somehow a light goes on (output). Pull the lever and the toilet flushes; turn a key and the car starts (usually). Not only in the big issues—having to do with the relationships among nations and the structure of the human experience writ large—but also in our everyday lives we have been surrendering our autonomy to things we comprehend only vaguely. Modern Homo sapiens lives increasingly removed from the primitive realities of rock and earth, water and wind, bird and leaf: things we can feel and understand primitively, if not intellectually.

In Greek mythology, the giant Antaeus drew his strength from the earth. He was invincible in combat so long as his feet were on the ground, but was eventually killed by Hercules, who strangled him while holding the hapless giant up in the air. We are also being lifted out of our element, and perhaps with similar results.

At one of Long Island's most exclusive country club beach resorts, ultra high heel shoes have long been de rigueur, worn daily by the socially active, jet-setting members. The most common injuries on the beach were not drownings or even abdominal cramps but rather, ruptured Achilles tendons in these high society women, suffered while walking barefoot on the beach. The requisite footwear for an advanced social life apparently caused shrinkage of the sturdy tendon leading to the heel. Hyperextending the heel, as a result of taking off the shoes and walking flat-footed in the soft, pockmarked sand, abruptly stretched the shrunken tendon and caused it to snap. Poetic justice, perhaps.

To the nonscientific mind, most things are either mystery or miracle. To the average twentieth-century inhabitant of any advanced Western nation, nothing has changed, except that we are expected to understand the mysteries and miracles on which we rely every day. Failing that understanding, emotionally uprooted by our dependence on things that are *from* us but not *of* us, our sense of connectedness is severely undercut.

Awakened by an alarm clock, he or she eats some packaged breakfast food and rides to work in a bus, train, or automobile. Once there our hero may sit behind a desk, conducting business over a telephone with people perhaps a thousand miles away, whom she doesn't even know. Or perhaps he performs some repetitious factory work, the end product of which he rarely sees. As tangible reward for such labors, there are pieces of paper that can be exchanged for things needed or wanted, the products of someone else's work. Small wonder that on Sunday afternoon he sets fire to some charcoal in his backyard and eagerly plunges his hands into a few pounds of raw hamburger meat, letting the blood squeeze out between his fingers; or she sways her body rhythmically and rapidly to high-amplitude sound, accompanied by large numbers of others, similarly disposed. At least this is real.

Not only has our culture placed us at a greater and greater distance from animal reality, also it subjects us daily to worries and stresses over which we have little if any control. Through television or newspapers we are informed of major political, military, social, and economic events which we feel individually powerless to influence, but which will clearly influence us. We are free to rage, worry, or approve, but the very size and complexity of our cultural apparatus makes it difficult for us to act effectively and see the fruition of such action.

During a recent eclipse of the sun, more people watched on television than in person, although it could have been viewed safely right outside. Shackles on the mind can be even more potent than on the body (or the Achilles tendon) and it is sobering to realize that culture is rendering us physically incapable of directly meeting the world, or mentally unwilling to do so.

Interestingly, outdoor activities such as hiking, canoeing, wildlife viewing, cross-country skiing, and mountain climbing are all excellent tonics for the complex culturally inspired disease of alienation. Such activities may provide their own kind of stress, but the stress is physical, direct, straightforward, and understandable. The simple realities of boot on ground, hand on rock, paddle in water—all are right there, without the interposition of bureaucracies, technologies, or ideologies. Small wonder interest in such activities has been growing exponentially in modern America.

And despite the intensive news coverage surrounding moon landings and space shuttles, children are still more likely to play cowboys and indians than astronauts. Not only has general interest in the space program fallen off rapidly—this was more or less expected—but the image of the astronaut has become remarkably dull and unexciting, while the cowboy rides on. The hero of our culture vs. the hero of our biology, and the latter is winning. The contrast is striking, the reason simple yet telling. Cowboys have a crucial something that astronauts lack: personal autonomy, direct physical control over events. The horse and gun, good guys and bad guys: these are simple realities, the stuff of cowboys and of cops and robbers and detective shows as well. And we long to make them our stuff too. Our fantasized spacemen such as Flash Gordon, Buzz Corbett, or Captain Video all participated in *personal* adventures in which strength, speed, courage, or intellect were successfully exercised in a straightforward manner.

By contrast, the reality of space travel is much nearer the reality of modern cultural-technological life. The erstwhile spaceman depends on an enormous support system, a "team" equipped with elaborate communicative, maintenance, and control devices. In fact, the astronaut is a virtual robot, pressing crucial buttons in response to orders from "Mission Control." If something goes wrong or a new course must be plotted, she must await orders from the computer and the engineers at home. The astronaut has virtually no real control over her environment, and aside from personal courage and technical competence, shows little that warrants emulation. As "doing" animals, we admire mastery over situations, while in her dependence on a myriad of imposing but somehow artificial cultural constructs, the astronaut reminds us too much of our own lives.

Greater mastery over space exploration is doubtless achieved by the concerted efforts of modern technology than could ever be accomplished by some rocket-mounted Buffalo Bill. But it is the culture's triumph, not the individual's. Our culture, in fact, has won many convincing victories. These triumphs, however, although not defeats for the individual, often leave us hollow and are sometimes Pyrrhic victories at best. For example, the institutions of mechanized agribusiness, combined with corporate distributorships, make food available to us in profusion and convenience. A great triumph of the "system." Yet what we gain in selection and facility we lose in satisfaction. It is rarely a very meaningful experience to buy some tomatoes at the supermarket. It's a different thing to feast upon a plump, red beauty you planted, tended, and watched grow. Not only that, but the home gardener has a known quantity; he has controlled its development. If for example, he doesn't want pesticides on his food, he can simply "make" it without them instead of depending on an unknown, unresponsive corporation, mass-producing a product perhaps thousands of miles away.

The technology of medicine has progressed similarly. There is a hollowness that accompanies many of our common practices, a feeling that the patient is being deprived not only of personal identity but of the right to meet life with the dignity of a sovereign being. Antidepressants and tranquilizers are prescribed with increasing frequency and unconcern, indicative of our preoccupation with symptoms rather than causes. We overmedicate nearly everything, with resulting natural selection for increasingly resistant disease organisms. Many women are still drugged into semiconsciousness during childbirth, admittedly sparing them some pain, but also cutting them off from the reality of one of the most intense experiences they can ever know. As with the increased interest in organic and home-grown foods, the growing popularity of "natural" childbirth reflects a growing rejection of the technologically mediated alienation doled out in modern obstetrical practice.

We are often rebuked for being a grossly materialistic culture, and yet in a sense, we are antimaterialist to an extreme. The philosopher Alan Watts once commented that a really materialist culture would have more respect for *material,* and would never tolerate the plastics, the veneer, the mass-produced throwaway garbage and shoddy workmanship that typically surround our lives. Similarly, although we are a nation of fat people—or at least, a nation of dieters—there is reason to doubt that we really value food. If we did, we wouldn't have consumed billions of McDonald's hamburgers. Instead, we seem to be captives of our own capabilities, impaled on our own hyperextended technology.

For example, one of the major innovations of the industrial revolution has been the factory. To understand its importance we must compare the theory behind factory operation with its more "primitive" predecessor, the artisan. When an individual's work is directed toward a whole, finished product, the activity can be deeply satisfying. Of course, the artisan must be well trained, and both apprenticeship and actual production are likely to be time consuming. Each worker in a factory, by contrast, specializes in a relatively simple operation, a very small part of the whole. Training takes less time and the product is completed more quickly since each step is handled by a different "specialist," usually with the aid of considerable mechanization as well. Job satisfaction is typically low. Potential satisfaction—the feeling of belonging and rapport not only with one's colleagues but also with one's work—has been sacrificed to efficiency and a sort of group mastery that can leave the individual isolated and unfulfilled.

On balance, and despite many attempts to break out of the situation, things appear to be getting more desperate as technology marinates in its own positive feedback. Holding a tiger by the tail, we fear to let go because we have traveled so far from the primitive state that we could not survive without our ferocious ally. Yet it carries us farther and farther into a land that is fundamentally strange, and therefore alienating as well as downright dangerous. And the more we struggle, the greater the alienation we produce, since our struggles necessarily involve the use of yet more artifice and artifact, which are basically the problem, not the solution.

Alvin Toffler's *Future Shock* describes another symptom of modern culture-induced alienation. A traumatic experience such as serious injury or the witnessing of some terrible event can result in "shock," a medical condition that may lead to death (see the discussion of stress-induced adrenal shock, earlier). Modern Homo sapiens—according to Toffler—is in a state of virtually continuous shock, produced by our inability to assimilate change at its new frenetic pace. We are disoriented, and it is getting worse as change comes faster and faster.

We are also an extraordinarily mobile society. It has become rare for children to make their adult lives near their birthplaces, and almost unheard of for families beyond the unit of husband-wife-and-children to live together. This likely deprives us of a great sense of continuity, to be found in living out our lives in the surroundings of our childhood memories. When people live in one place for a long time, they are said to have "put down roots." A plant gets nourishment and strength from its roots, and a sense of permanence and relatedness to the earth. By contrast, modern Western Homo sapiens lives in a state of seemingly permanent transience, rarely living in—or even remembering—our childhood homes, changing jobs and friends, schools and doctors, seeing only one very narrow side (the work-related one) of perhaps 99 percent of the people we meet. (In a term that never caught on, but deserved to, Toffler suggested that most working relationships were of such brief duration that instead of bureaucracy, we should identify "ad-hocracy.")

To some extent the seeming perpetual motion of modern Homo sapiens is a direct result of our cultural system. We go where the jobs are, where we are transferred, or in times past, where the draft could be avoided. In such cases, we seem to be helplessly dragged about by the riptide of events our culture has created. On the other hand, much of our mobility is intentional. Culture has given us the opportunity to better our lot. We can actively seek a better job, more pleasing climate, specialized schooling or medical care; mobility may well be a valuable tool for widening our horizons and making us happier. But regardless of whether our rootlessness is forced upon us or comes as a result of free selection from a newly opened range of possibilities, it is rootlessness nonetheless, and its effect on our psyche remains the same: we become strangers to our own land, our fellows, and to ourselves.

As a result of this estrangement, many people are also likely to feel estranged from society's customs, norms, and expectations. Earlier, we developed the notion that if Homo sapiens has a biologically predisposed preference for a particular kind of social structuring, such a preference is likely to be frustrated by the great multiplicity of human societies. We also suggested a more likely prospect, that our species lacks such a preference, and that as a result, we are also subject to disruption of our cultural systems and feeling alienated from them, since we lack the animal's innate patterns of social adherence, acceptance, and stability.

Let us assume that the human species, despite its rigid physical body, is basically amoebalike, able to ooze into almost any culturally defined shape. The role of culture, then, would be crucial in stabilizing our fundamental formlessness, providing structure and direction to what would otherwise be confusion and chaos. If, in the beginning, human nature is without form, and void, and darkness is upon the face of our being, then we owe a great debt to the Spirit of Culture, which moves over the face of the waters and says, "Let there be this way of living, and that." And "Thou shalt do this, and not that." In a sense, the genesis of human culture therefore unburdens us of the options of too many choices and too little human nature.

What happens, then, when alienation increases to the point of undermining confidence in one's culture? Unprecedented freedom can become confusing, even paralyzing. Independence can be transformed into fragmentation and disorientation. Individuals can find themselves wayward, lost, angry, and unfulfilled, uncertain what to choose for their lives, or even how to formulate such choices. When the past seems unacceptable as a guide and the future unpromising, the present assumes an overwhelming weight, which it cannot bear without collapsing.

We have a physical structure, our skeleton; in other respects, however, we cry out for structure, and yet our culture often provides us with structures that are profoundly inadequate and alienating, because they are too rigid, too uncaring, or too downright dangerous. Increasingly alienated from their cultures, people turn to other cultural practices that promise relief: primitive religion, violent terrorism, psychoanalysis, sociopathy, or involuted indifference.

In the years before Sigmund Freud, psychiatrists were commonly known as "alienists." And not surprisingly, the stuff of alienation is still the stuff of mental illness. As Freud pointed out in his *Civilization and Its Discontents,* civilization requires to some degree the frustration of basic human tendencies: aggressiveness, cruelty, violence, sexual debauchery, excessively selfish individualism of all sorts must be constrained if human beings are to live amicably with each other.

Just as civilizations can mimic and exaggerate biological traits through the process we have labeled cultural hyperextension, they are also prone to overreact to perceived human tendencies, tendencies that may be either lacking altogether or actually benign, but pathologic when frustrated. For example, in his book *Touching: The Human Significance of the Skin,* anthropologist Ashley Montagu described the various pernicious ways in which human societies have denied the natural, harmless, and healthy human need of children and adults for touching and being touched. The child's inclination for self-expression often leads to repressive school discipline; sexual yearnings lead to Victorian taboos, and so forth, all combining to produce a human being who is at best alienated and neurotic, and in whom any vestiges of mental health are testimony to an extraordinary constitutional toughness.

Radical psychiatrist R. D. Laing may have been exaggerating only a little when he wrote, in *The Politics of Experience:* "From the moment of birth, when the Stone Age baby confronts the twentieth century mother, the baby is subjected to these forces of violence, called love, as its mother and father, and their parents before them, have been. These forces are mainly concerned with destroying most of its potentialities, and on the whole this enterprise is successful."

To theologian-author Andrew Bard Schmookler, this frustration is specifically designed to produce rage and lust for power: "Power demands better servants than human beings are, as nature created them. Civilized societies need power, and thus they are compelled to re-create man. Socialization practices are the instruments by which social demands are translated into psychological structure." His explanation, though ingenious, seems overly contrived. More likely, the anger and frustration that follow from the practice of socialization are simply a result of the discordance between biological and cultural factors. Sometimes, our cultural practices demand too much of human beings; sometimes too little. Only very rarely can we expect them to hit it (that is, us) just right.

The strains generated by conflict between Christian theologic doctrine and biologic inclination were well portrayed by Saint Paul, when he wrote in anguish: "I delight in the law of God after the inward man: but I see another law in my members, warring against the law of my mind, and bringing me into the captivity to the law of sin which is in my members. O wretched man that I am!" (Romans 7: 22–24). Or as Hamlet might have said, O cursed spite, that ever culture tries to make biology right.

We began this chapter by looking at alienation as something that results from our culturally created environments, and have subsequently been moving toward examining alienation as something that also results from our unique consciousness, a level of self-awareness that apparently exists only in Homo sapiens. Apes and chimpanzees have a "self-image," or at least they will seek to wipe off grease paint applied to their faces once they see themselves in a mirror. But no animal can look at itself and be sad or disappointed. No animal obsesses about the gulf between the way things could be and the way they are. No animal is aware of its own, unavoidable death, and yet at the same time, aware that nothing can be done about it. Animals may be sad—sometimes downright miserable—but no animal feels sorry for itself, or feels worried about the fact that it is sad or disconnected.

Insofar as there are certain "inalienable" aspects to human nature, the more culture and human nature fail to correspond, the more strain will be produced, and the more alienated will be the human products. Eventually one or the other may crack: either the offending cultures will then change, or the offended individuals will develop neuroses or even psychoses.

Modern ecologists, Eastern mystics, and drug-stimulated

Westerners have recognized, apparently independently, that organism and environment are not separable, that the human skin, for example, does not isolate us from the rest of the world but rather, joins us to it. Flying in the face of this recognition, however, is the alienation of Homo sapiens from the manmade environment, from other human beings, and sometimes, from himself. A brilliant and sensitive man—also a former mental patient—Charles M. S. Sade (no relative to the notorious Marquis de) describes the experience of schizophrenia as resulting from a profound alienation between the biological self and the cultural whole. Sade writes that the schizophrenic

> has chosen himself, his limited individuality, his ego, to be his whole world. He usually seeks himself in a gradually enlarging interiority of mind. He is self-conscious, brooding, introspective, often passive-dependent externally, often hearing voices, which convey subtle and sensitive condemning or glorifying meanings. I don't think his ego is split or gone; in fact, it is probably magnified with delusions of grandeur which compensate for the real loss, the real split from his truer self that is being at one with things, people, and the world. To the extent possible he becomes an isolated fragment, alone in a mind, often tormented by guilt for his refusal to accept the larger reality. He nurses his secret hurt and in defiant pride, to affirm his isolated self, projects a solipsistic metaphysic, seeing the spirits of his inner world as ministering either to or against him and him alone, while what is left of the outer world means only in relation to him. Since he cannot actually completely withdraw from the outer world, though he seeks it, it impinges upon him as wholly other, mostly negative, blind to him or totally against him. So in some little place he has found to sit he asserts his kingship as the most important, most deeply persecuted person. His body means little to him and like a burden it weighs him down. He becomes finally a lonely, completely impotent god.

The loneliness of the schizophrenic is the loneliness of the human being isolated from his or her legitimate connections. The impotent god is transfixed by the disparity between his biological creatureliness and his culturally mediated alienation—often with an assist from some chemical imbalance in his very biological brain.

It is unclear whether the frequency or severity of mental illness is any greater in modern times. However, it seems likely that during the past few decades there have been significant increases in the use of mind-altering drugs, accompanied by a legitimate and growing concern on the part of parents, educators, and law enforcement people, who unfortunately are often terribly misinformed as to their effects and causes. Psychedelic drugs provide both escape from reality and a feeling of enhanced experience. ("Reality is only a crutch," reads a bumper sticker, "for people who can't handle drugs.") It does not really matter for our purposes whether these experiences are imagined or real, that is, whether one's sensitivity and perception are actually heightened or whether seeming revelations are just hallucinations, imaginary perceptions carrying no true insight into underlying realities. The important thing is that it seems that way to the user.

Use of marijuana, LSD, and more recently, cocaine, has become particularly prevalent among middle- and upper-class youth. In fact, it was the explosive increase of use among these groups that finally prompted today's massive social concern and even the rumblings of political action. These drugs have gained a wide audience because of the enhancement of perceptions and feelings they offer. They are not particularly new discoveries. They are not the latest wonder drugs, or recent contributions to "better living through chemistry." Marijuana and mescaline have been around for a very long time, although perhaps never achieving the popularity and the notoriety they enjoy today. They have been rediscovered overnight, by a populace weary of their alienated, secondhand approach to life and eager to be turned on and tuned into the vibrant sensations of a reality of which they sense they are being cheated. Drugs are a means to an end.

Actually, despite the possible health hazards involved, the use of mind-expanding drugs may be one of society's least dangerous efforts at overcoming alienation. For some, it appears that a true sense of living can only be achieved through acts of violence, most notably the killing of someone else. "Thrill murders" and serial killings are frequently perpetrated by deeply alienated people whose psychotic boredom fairly screams for behavior of high intensity, in order to make them know they are alive. While such behavior is clearly pathological, similar feelings, although less extreme, underlie many common features of everyday behavior. Snowmobiles, downhill skiing, driving cars, boats, and motorcycles too fast: all these are socially acceptable means of declaring a measure of personal mastery and autonomy. By seeking thrills with a degree of danger, technological man seeks to insert himself back into the world, to feel once again the pulse of real, biological life.

The need to feel such life seems especially acute for a species—the only one, as far as we can tell—that is aware of its eventual death. Awareness of death can add a certain spice to life, or it can drown existence in hopelessness and a deep, existential alienation. It can also lead to religions and cults, in a primitive search to achieve via cultural constructs the pseudo-community and secure rootedness of which knowledge of death has deprived our biological selves. If we were intelligent computers instead of human beings, then perhaps awareness of corrosion, short circuits, and the possibility of power outages would lead to an existential electric angst and along with it, a deep-seated alienation from our electronic selves and the disadvantages and perils that necessarily follow from this fact. Instead, we are intelligent animals, and therefore doomed to being aware of the limitations to our creaturely existence, limitations that are ultimately unavoidable even as we may perceive them to be tragic. We may elect to go gentle into that good night, or to rage against the dying of the light—but one way or the other, go we must.

In Plato's *Phaedo,* Socrates defined the fundamental wisdom of philosophy as "being mindful of death." He did not say, however, that we had to like it. On the one hand, mindfulness could lead to a fuller existence, if not to happiness. On the other, mindfulness in biological creatures cannot but lead to a painful awareness of our predicament, circumscribed as we are by our organic bodies and the certainty of our eventual death. The fundamental alienation of a creature doomed to die is thus a reassertion of our inescapable biology.

As Arthur Koestler emphasized, we are not "programmed" to conceive of our own deaths. When a computer confronts something for which it is not programmed,

> it is either reduced to silence, or it goes haywire. The latter seems to have happened, with distressing repetitiveness, in the most varied cultures. Faced with the intractable paradox of consciousness emerging from the pre-natal void and drowning in the post-mortem darkness, their minds went haywire and populated the air with the ghosts of the departed, gods, angels and devils, until the atmosphere became saturated with invisible presences which at best were capricious and unpredictable, but mostly malevolent and vengeful. They had to be worshipped, cajoled and placated by elaborately cruel rituals, including human sacrifice, Holy Wars and the burning of heretics.

Alienated from the facts of our lives, we perpetrated death on others.

Within eight years of writing these words, the brilliant Koestler, nearly eighty years old and terminally ill, took his own life.

CHAPTER THIRTEEN

THE BIOLOGICAL FUTURE

Advice from the Cheshire Cat

The horseman serves the horse,
The neatherd serves the neat,
The merchant serves his purse,
The eater serves his meat;
'Tis the day of the chattel,
Web to weave, and corn to grind
Things are in the saddle,
And ride mankind.
—Ralph Waldo Emerson

And what does biological evolution have in store for us dual creatures? Will we ride our culture or vice versa? It is inconceivable that we shall become in any way less biological in our nature and yet, equally unlikely that culture will somehow exert a lesser influence than it does today. In fact, despite occasional minority pushes for simplification in life style, we can expect a continuing increase in the technologic, abiologic aspects of human existence. Of course, we will also remain subject to natural selection. So long as some people have more offspring than others (for whatever reason), selection will occur, with the population's gene pool shifting in their direction.

But not only will this selection be—as usual—too slow to

counteract or even effectively harmonize with cultural evolution, it will also become increasingly less "natural." As birth control information becomes more widely disseminated, the major factor controlling reproductive success will be the rational intent of the prospective parents. Religious affinities, ethical and moral judgments, financial considerations, political and social atmosphere, and a host of personal factors will be the ultimate criteria, so that the evolutionary fitness of each individual and family will probably have very little genetic basis. Even the evolutionary fitness of genes will have little relationship to the characteristics of the genes themselves. This conscious, extragenetic influence on biological selection is unique in the history of evolution and there is no predicting where it will lead.

But it is nothing new for our species. "There is no stopping place in this life," observed the German theologian Meister Eckhart nearly 700 years ago, "no, nor was there ever one for any man, no matter how far along his way he'd gone. This above all, then, be ready at all times for the gifts of God, and always for new ones." We can substitute cultural evolution for God, if we wish, but the message remains. Thanks in part to nuclear weapons, environmental pollution, resource depletion, and overpopulation, the future isn't what it used to be. But it still is.

In the twentieth century, Alfred North Whitehead observed that it has always been the business of the future to be dangerous. Maybe so. Never before, however, has the future been doing its job quite so assiduously.

Insofar as biology goes, the human evolutionary future, in fact, is neither completely obscure nor demoralizing. We can assume that Homo sapiens will carry an increased load of deleterious genes (i.e., become somewhat "inferior") and will therefore come to rely increasingly upon supportive technology. This seemingly glum prophecy is not really so dire, or even so undesirable, as it might sound, however. It simply reflects our growing proficiency in the healing arts. To take one example, sufferers from diabetes had little hope just one hundred years ago; today, with artificially administered insulin, they can live virtually normal lives. Without medical treatment, diabetics are much less likely to have offspring than are healthy individuals; with it, there may be no difference in reproductive rates.

Since diabetes is genetically influenced, we can expect an increase in the frequency of associated genes, the natural selection pressure against them having been relaxed by an important cultural practice. Of course, such an increase in diabetes is less than catastrophic, since we have the capability to control the disease; among diabetics and their relatives, the merits of insulin injection can hardly be questioned. (Someday soon, we shall probably have the technology for pancreas transplants.) Ethically, we have no choice but to persist with cultural developments of this sort, even though they force the human species higher and higher up a slippery pyramid of dependence on cultural evolution. It is often easier to climb up than to get safely back down.

Among our biological characteristics, eyesight is a good example of the modification of "natural" selection by cultural evolution. Good vision must have been extremely important among our ancestors, as it still is for nontechnologic people today. It is obviously desirable to spot prey at a distance, ideally, before they can spot you, as well as to be aware of predators. Individuals with severe myopia would be at a selective disadvantage, and we might expect genetic combinations for good vision to be maintained at a high level in populations exposed to purely biological evolution. But, thanks to the optician's art, 20-20 vision is now bestowed upon anybody who can afford it, granting equal chances of survival and reproduction regardless of genetic constitution. When it comes to evolutionary change, culture has become the great equalizer. Even if culture was somehow unable to correct artificially for genetic inequalities in vision, other aspects of our culture have rendered visual acuity increasingly less important. Thus, possibly aside from increased danger of being hit by a car or walking into an open manhole, the visually incompetent in modern society would probably experience no severe reproductive difficulties.

The necessities of survival and reproduction are available for most Americans, and for many human beings the world over. "Superior," fortunate, or simply ruthless persons may often avail themselves of more comforts, luxuries, and satisfactions, but at least in America it is unlikely that they will have larger families. In fact, the reverse may be true. The "right" of all people to have children is not at question here: like all other rights, it is granted by society and is presumably beneficial to society as a whole. Nonetheless, our culture is tampering with the fundamentals of our biological evolution, and we should be prepared for the consequences, or rather, for the fact that there will be consequences, even if we cannot now anticipate them.

We were practicing *artificial* selection for thousands of years, long before the genetic basis of its success was understood. We have created distinct strains of dogs, for example, by choosing those individuals with characteristics we wanted, and causing them to breed with each other while prohibiting "undesirable" matings. By thus interceding between our domestic animals and natural selection, we have produced such divergent forms as the Chihuahua and the St. Bernard. From cow to chicken, our common farm animals also differ greatly from their wild relatives, upon whom only biological evolution had operated. Nearly all our food plants are likewise the result of artificial selection, by which we have "designed" strains with higher yields, greater disease resistance, and so forth. There is nothing new, then, in artificial selection; when it comes to biological evolution, Homo sapiens is in the habit of playing god.

But although our domestic plants and animals are "better" (for our purposes) than their wild relatives, they are also generally inferior to them in direct competition. For example, compare the barnyard pig with its probable ancestor, the wild boar. The farmer's pig has comparatively dull senses: keen vision, hearing, and smell are not of particular selective value in the barnyard environment, while they are clearly important for the free-living boar, a vibrant, alert, and energetic animal. The pig is slow moving and lethargic, a selective advantage to the farmer, who prefers his animals to gain weight rather than waste a lot of calories running around. The contrast between the domestic and the free-living is striking.

There may be a parallel between the evolutionary changes brought about by domestication and those potentially induced in ourselves by civilization. In both cases, biological entities are isolated from the rigorous pruning of natural selection. Instead, human beings have become the gardeners—and in our gardens, we cultivate ourselves.

In popular stereotype, the future human being is pictured with an enormous head perched upon a feeble body. This is a misconception. It is probably based on unconscious Lamarckian thinking. Under this view, traits acquired during a lifetime are passed on to one's children. Thus, it is reasoned that since we seem to use our brains more and more—reflecting the increased pace of cultural evolution—our successors will somehow be born with larger and larger heads. For this to happen, however, brains (like muscles) would have to get larger when they are used. Moreover, there would have to be a selective advantage to large brains, as there was when culture was first developing. In other words, for biological evolution to proceed in this way, people with greater intelligence would have to be producing more children than those with less intelligence.

As mentioned above, there is some reason to believe that if anything, the opposite may be occurring, with culture now operating to make possible equal genetic contributions from all, regardless of the biological qualities of each. There is even some potential for modern cultural evolution to reverse our earlier biological trend toward increased brain size. Thus, enlightened individuals with greater access to contraception and greater sensitivity to its importance are probably more likely to have smaller families.

Actually, we must carefully refrain from judgments of "superior" or "inferior" in this regard, since for biological evolution these terms simply mean the capacity to produce successful offspring and other carriers of one's genes. As far as evolution goes, lower intelligence may well have become superior. The distinction between biological and cultural values is underlined by the fact that Leonardo, Newton, Beethoven, George Washington, and Jesus Christ were all failures of natural selection, since none produced any offspring.

The situation is actually cloudier yet. Although it is fairly safe to say that socioeconomic level correlates inversely with family size—wealthier people generally have fewer children—there is virtually no hard evidence showing a consistent correlation between socioeconomic level and intelligence, and demographic trends are notoriously variable. The Kennedy mystique made large families fashionable; ecological consciousness caused a reversal; no one knows what will come next. Also, it is not necessarily true that brain size and intelligence are very closely related, at least within the current normal human range. Albert Einstein had an unusually small brain.

Although selective pressures on the human population have changed during the course of cultural evolution, so that we are exposed less and less to selection based on "traditional" biological qualities, natural selection is still occurring; only the framework has changed. One thousand years hence (or maybe even one hundred) biological evolution may be selecting for abilities to withstand the new pressures of human existence. Just as the use of DDT selected for resistant flies by killing off the vulnerable ones, increasing air and water pollution may select for people with inherent resistance to various contaminants. Just as the discovery of insulin has slowed down selection against diabetes, maybe increased smog levels will select for greater respiratory resistance, by killing off people with a tendency toward emphysema or lung cancer. (It is said that there are only good drivers in New York City; all the bad ones have been killed off.)

When an organism is placed in a contaminated environment, selection favors those individuals that can live and reproduce in spite of it. Perhaps the future human beings will be those who can live and reproduce despite their new internal environment of lead in their blood, mercury in their brains, strontium-90 in their bones, and PCBs in their fat. Beyond this, cultural evolution is providing a whole gamut of new, rather subtle environmental changes, which may also exert novel selective effects on Homo sapiens in the future. Our successors, for example, will have to cope with high noise levels, crowding, and the increased pace and attendant strain of modern technological life.

While this analysis may sound logically correct, it could equally well be all wrong. We are talking about many years and many generations. Given the present rate of cultural evolution, no one can predict the human environment just a single generation from now, never mind the hundreds required for substantial biological change. Modern physical contaminants and social environments may be entirely gone in a century or so, replaced by a whole new set of culturally generated conditions. Organisms in nature eventually reach a sort of equilibrium with their environment, so long as the environment itself remains more or less constant. But if the surroundings won't hold still, there is little chance for the living things to catch up. This may happen to us.

It is curiously satisfying to imagine a future in which biological evolution has somehow attuned us to the productions of our culture. But it is unlikely. Not only will culture always outstrip biology, but strangely enough, our own physical hardiness and short-term adaptability must ultimately get in the way of a lasting biological accommodation. Natural selection is generally concerned with an organism only *until* it has reproduced. This is why most living things do not live very long after breeding: having produced their offspring, and nursed and weaned them to independence, evolution is no longer very much interested in them. More accurately, deleterious mutations would quickly be selected against if they expressed themselves in a way that interfered with fitness; after one's genes have been sent on their way, the fate of the progenitors exerts relatively little effect, any more than a rocket designer is interested in the fate of the booster stage after the satellite has been launched.

Human beings, possibly whales and elephants, and perhaps some of the other primates are exceptions. They experience a fairly lengthy postreproductive life span because among animals for whom learning and judgment are important, the greater experience of older individuals is of selective advantage. With the invention of writing, the printing press, microfilm, and computer data processing, the value to society of elderly and knowledgeable people might seem to have progressively declined. If the pace of cultural change continues to accelerate, causing experience to be a detriment insofar as it is rooted in a different and hence, irrelevant past, we could also expect this trend to accelerate as well. (It seems noteworthy that this potential change in the usefulness of the elderly may be more true for technology than for nontechnical aspects of culture, such as religion, diplomacy, history, and so forth. And even with regard to the latest in path-breaking technology, graybeards may be of special value, in helping us recognize how new developments depart from previous experience, how on the other hand such developments may be simply reinventing the wheel, and/or how they may offer dangers as well as promises, based on similar events in the past.)

In any case, for natural selection to be effective in biologically adapting the human population to its culture, our new environments must somehow influence reproductive performance. This in itself is unlikely, since we are sufficiently sturdy that most culturally induced disease and illness do not seriously affect us until middle or old age. Thus, although acute mercury and lead poisoning may strike young children, most of the stresses produced by cultural evolution result in so-called degenerative diseases that have little influence on reproduction. Such increasingly important illnesses as heart disease, emphysema, and cancer seem to be caused by an accumulation of environmental "insults" like stress, anxiety, and a whole range of irritating and dangerous substances, chemicals, and situations. Since they generally don't affect us until after we've reproduced, however, there is little opportunity for those with greater resistance to leave more offspring. It is therefore unlikely that we could evolve much resistance even if the rate of cultural change was drastically slowed down.

But even if evolution doesn't care about the increased prominence of degenerative disease, we should. As part of the higher aspects of our mental evolution, we have developed a profound sensitivity to the survival and suffering of others. Perhaps this is an outgrowth of kin selection or reciprocity, or perhaps it is a purely cultural creation. In any case, we have reason to care, if only because we can easily imagine ourselves in the situation of others, and this capacity of anticipating the future is by itself another uniquely human attribute. By their nature, however, degenerative diseases build up slowly and often they are not clearly associated in the public mind with the environmental factors that are actually to blame. This makes alleviation of the problem yet more difficult.

Heart disease is probably our most serious degenerative illness; it has achieved almost epidemic proportions in the United States. This example is particularly interesting since it relates to the rest of the big-brained stereotype, the deteriorating body, while it also provides one of the most striking examples of the conflict between biological and cultural evolution. Our frighteningly high incidence of heart disease results especially from four combined factors: stress, eating habits, lack of exercise, and the use of certain drugs, notably tobacco and alcohol. All four have important biocultural aspects. We have already considered the likely contribution of cultural evolution to stress. Let us now consider the remaining three.

There is no doubt that being overweight is hard on one's heart. Our eating habits leave much to be desired. We seem to like things that aren't good for us. Why?

Here is a possible scenario:

Our ancestors probably ate lots of fruit. Modern primates still do. Fruit is nutritious and also high in sugar. It therefore became selectively advantageous for primates to like such foods, so enjoyment of sweets became part of our biological makeup; that is, individuals with such preferences left more successful descendants. As cultural evolution progressed, we began creating our own culinary environments, instead of simply depending on whatever nature had provided. We processed our own food, and, given our primate predisposition for sweets, we learned to make a whole array of "taste-tempting" delights. These tend to be very high in sugar and carbohydrates, and often little else. Human cultural evolution has thus taken a basically adaptive biological characteristic and perverted it, much to our immediate pleasure and ultimate detriment. Like our preference for big-headed cartoon children and big-busted silicone women, our sweet tooth is another culturally exploited supernormal releaser, a form of cultural hyperextension that has been paying special dividends to the confectionery and mortuary industries, and more recently, to physical fitness parlors.*

It seems "natural" for us to like sweet things, and indeed it is, just as it is natural for a raccoon to like crayfish or an anteater, ants. Our fondness for the taste of sugar is predicated upon its typical occurrence in nutritious foods present in the early human environment. Interestingly, the chemical lead acetate also tastes sweet to us, but is a deadly poison. Lead acetate is not normally found in our environment, and never was, certainly not in any abundance. If it had been, we would have evolved the ability to distinguish its taste from that of sugar, or we would never have considered both to be "sweet," or we would have gone extinct long ago.

In addition to sweets, we are also victimized by our fondness for "rich" foods, high in cholesterol. And this may be another ex-

*In *Sweetness and Power,* anthropologist Sidney Mintz has recently detailed how our biologically influenced fondness for sugar has become interwoven with a wide array of complex cultural practices, including slavery, imperialism, power politics, and industrialization.

ample of supernormal releasers and cultural hyperextension, due this time to the carnivorous habits of our Australopithecine ancestors. Wild game is usually quite lean, and animal fat, with its high energy value, was therefore undoubtedly prized by early hunters, as it still is among nontechnologic people. Our domestic animals currently produce enormous quantities of fat, and the beef industry gives special attention to well-marbled meat, which we find tasty and highly desirable. Unless restrained by new-found cholesterol consciousness, we gobble it up even as it lethally clogs our arteries.

Insufficient exercise is another prong on our trident of factors causing high rates of heart disease. To see its relevance to our theme, we must go back once again to the African savannah. The world in which human beings evolved was intensely physical. We had to run from our enemies, run to catch prey, walk long distances for water or shelter. Our physiology and anatomy, in particular our circulatory system, evolved with the expectation of vigorous stimulation for growth and proper maintenance. But although exercise may have become necessary for our health, we didn't have to like it.

One of the most notable achievements of cultural evolution has been the steady replacement of human labor by machines. Transportation has been particularly revolutionized, to the point that although an Australian aborigine may think nothing of covering twenty miles a day on foot, the average American would be horrified at the thought of walking a mile or two. We do almost all our traveling on our backsides, and our electrocardiograms have been showing it.

But if we evolved a fondness for sweet things "because" they were good for us, why didn't we also evolve a natural fondness for exercise, which is equally beneficial? Perhaps physical exercise—unlike ripe fruit—was unavoidable for our ancestors, so there was no need to generate a predisposition for it. Furthermore, it was certainly advantageous for the early hominids to take shortcuts and use cooperation or primitive tools to save energy wherever possible. We may therefore have evolved both a need for exercise and a tendency to be lazy.

Finally, what about tobacco and alcohol? Human beings are stimulus-hungry animals, and not only for sights and ideas but for things that we eat, drink, and breathe as well. Even today, perhaps the best assurance of a nutritionally adequate diet is one that is varied. We possess—and sometimes we appear to be possessed by—a powerful drive to stimulate our taste, smell, and other senses with sensations including those provided by alcohol, tobacco, and spices. It is not widely appreciated that prior to 1492, smoking tobacco was unknown to the rest of Homo sapiens, and that what we might call the Red Man's Revenge has been quite spectacular if not complete: it is almost certain that however many Native Americans were slaughtered directly by the invading Caucasians, and indirectly by their alcohol, even more Caucasians have been killed by tobacco in the years following.*

Freud once referred to the human species as a prosthetic god. With culture hyperextending that prosthesis, we are able to employ our artificial limbs to great effect. But unlike gods, we can also get "stressed out," paunchy, flabby, atherosclerotic, hypertensive, and we smoke and drink too much.

If biological evolution did occur by the inheritance of acquired characteristics, our proverbial weak-bodied human would undoubtedly be the *Homo* of the future. However, there is presently no evidence that weak-bodied people leave more successful offspring than do sturdier individuals; evolution of this sort by natural selection is therefore highly unlikely. So the best we can say is that whereas culture is undoubtedly having an enormous influence on our current biology, the shape of our future—and indeed, of *us,* in that future—is virtually unknowable.

Genetic diversity is essential to biological evolution and to the survival of ecological systems. When a new environment provides

*Poetic justice, perhaps, like the observation that in North America, the buffalo are coming back, while the railroads are going extinct!

new situations, natural selection chooses the best from among the available variety. The greater the variety, the greater the chance of good adaptation, and the lesser the danger of extinction. And yet, cultural evolution appears to be drastically reducing the worldwide biological and cultural diversity, thereby jeopardizing the future.

The last few hundred years have been a time of cultural homogenization, as diverse human societies—generally "primitive" in their technology—have disappeared, to be replaced by imitations of Western culture. Visit any of the world's large cities: the languages may differ but the life styles are essentially the same. The same clothing is worn in Nairobi, Bogotá, New York, Tokyo, and Brussels. When a human culture is gone, it is virtually lost forever, almost like a species that has gone extinct.

One of the dangers of a "monoculture" such as a cornfield is that it is unstable, and at the mercy of environmental change. Cultivated monocultures may do beautifully for a while, given constant human attention, but when a pest such as corn blight comes along, the whole crop can be wiped out. If all the corn plants are from the same strain, as they frequently are in modern "scientific" agriculture, there is no diversity to provide a reservoir of resistant individuals. In addition, healthy diverse systems have a self-damping effect on perturbations, analogous to negative feedback. Epidemics, for example, do not readily spread when susceptible individuals are surrounded by others in whom the disease cannot take hold. The parallel to human societies is not mere analogy.

In natural communities, each organism may produce descendants, and not only individuals but species as well thus tend to replace themselves sequentially. Imagine leaves in a deciduous forest during autumn: each falls separately, and is replaced in turn. To be sure, there are connections and crucial interdependencies, but also enough separation that failure for one is not disaster for all. With the advent of modern worldwide culture, however, we have not only homogenization but a tendency toward convergence as well, as though the fate of all the leaves in a forest was depending on the success of just one tree. The world, in short, has become very tightly wired: what happens in one place affects others. Court intrigues among the Incas 1500 years ago had virtually no effect on the simultaneous goings-on at Camelot. But now, when Beirut, Moscow, or Tokyo sneezes, someone on the other side of the world is bound to say *Gesundheit.*

Different human societies have different requirements, survive in different ways, and contain differing reservoirs of competence in dealing with the world's realities. With only one "world culture," we would be greatly limited in our ways of making a living. With diversity wiped smooth, we would all be equally susceptible to environmental perturbations such as epidemics or resource shortages. Furthermore, as local cultures go extinct in favor of the highly technological society of the West, all of the world's people come to rely more and more upon technology's elaborate superstructure. This may cause a temporary improvement in living standards, but at a dangerous cost, not only in the aesthetics of human variety, environmental abuse, alienation, and resource consumption, but also in social flexibility and adaptability.

Aside from all the personal, rootless insecurity that comes from a loss of cultural identity, such rapid change often transforms a society that had been coexisting harmoniously with its environment into one that is antagonistic, no longer fitting in with its land or each other. Furthermore, excessive worldwide reliance on a complex technology places the human species at the mercy of breakdowns in that technology.* Imagine the Eskimo, having abandoned his dogsled generations ago in favor of the snowmobile, who suddenly finds parts and fuel unavailable, for any of a variety of reasons. The dogsled required only what the arctic environment naturally provided: wood and sinews for the sled itself, fish or caribou meat to feed the dogs.

*A cartoon once depicted two Eskimos turning to each other, as the nuclear missiles arched back and forth overhead, saying, "Well, that's the end of civilization as *they* know it." Increasingly, it is civilization as the Eskimos know it, too.

Given their native cultures, representatives of the human species in the far north, for example, could conceivably survive a wide range of cataclysms, perhaps even nuclear holocaust itself. But once they become reliant upon a single culture, the fate of the entire human species becomes concentrated into a very few hands. Even now, humanity is composed of a diversity of people, living in different ways, in places ranging from deserts to tropical rain forests, to mountain meadows, ocean shores, and the Arctic. If we contain the nuclear Neanderthals among and within us, we stand a reasonable chance that some of us at least will experience a long evolutionary history. But as we become a homogenized monoculture, we put all our eggs in one basket, and a fragile one at that.

The danger in reducing human diversity goes beyond our destruction of indigenous cultures. It extends to the more fundamental genetic level of biological characteristics as well. Thus, each human racial group possesses certain distinctive genetic combinations, some of which are ultimately responsible for our recognizing the different races in the first place. Although there is no basis whatever for comparing the different races on any global scale of relative quality, it is certainly possible that each race is better adapted than any other to its particular environment, because at least in the past, natural selection has operated on the inhabitants of each given region, adjusting the surviving occupants to local conditions and vagaries. It is often difficult to identify the specific advantages of each characteristic, but the black African's skin is probably an adaptation preventing dangerous sunburn and aiding the radiation of excess heat, while the very pale Scandinavian complexion may be a reverse adaptation, enabling the cold-climate inhabitant to gain maximum advantage from the limited available sunshine. The short, squat Eskimos are well adapted to conserve body heat in their cold environment, and the tall Zulus to radiate it.

Isolated human populations often possess distinct frequencies of blood types and also, differences in disease resistance. The

American Indians and Eskimos were decimated by pneumonia and tuberculosis, to which they had little resistance. Similarly with the native Hawaiians and measles, a disease that was rarely fatal to the Western missionaries who brought it as a bonus along with the New Testament. When Charles Darwin made his famous voyage around the world, collecting much of the information he would later use to buttress his theory of evolution by natural selection, he visited Tierra del Fuego. These bleak and stormswept islands at the extreme southern tip of South America present a formidable environment for human beings. Yet they were inhabited by two tribes, the Onas and the Yaghans, who lived very simple lives of fishing, hunting, and gathering, entirely isolated from the "civilized" world of the nineteenth century. Today, both groups have been civilized to extinction.

To the hard-hearted objectivist, with no more empathy than knowledge of evolution, this may even seem like a good thing, resulting in "improvement" of the species. But if anything, the opposite is true. Although we have thus far considered local extinction as the elimination of "unfit" types (e.g., the demise of a people *lacking* certain disease resistance), these people may have also possessed various unique genetic attributes. Potentially "desirable" traits are lost forever along with their carriers. The Onas and Yaghans, for example, had an almost superhuman resistance to cold and physical privation, enabling them to live near-naked in virtually constant thirty-degree wind and sleet. But we shall never know how they managed it.

Although there is a particular poignancy about the loss of certain human biological traits, such as the Fuegians' cold tolerance, other vestiges of our biology are less appealing. Sickle cell anemia, for example, is a serious genetically transmitted disease of American and African blacks. It is caused by a double dose of a gene that is prevalent in equatorial West Africa, among other places, the original homeland of most of the American black population. The disease causes malformation of the blood cells ("sickling"), clogging small blood vessels and causing great pain and debility, while interfering with the transport of oxygen to the body tissues. Considering the severity of sickle cell anemia, one might expect natural selection to have eliminated it long ago from the African gene pool. But the gene has flourished, sometimes reaching a frequency as high as 40 percent. This is because a single dose of the sickling gene causes partial immunity to malaria, which is frequent where the sickling gene also prospers. Thus, despite its strong disadvantage in double dose, its advantage in single dose has kept the gene around. Once malaria is eliminated or adequate preventive drugs are available, or the people afflicted live somewhere else, there is no more "saving grace" to sickle cell anemia. There is no ethical argument for saving the sickle cell gene. But it should at least be clear that few things in nature are all bad or all good.

"Would you tell me, please, which way I ought to go from here?," Alice, lost in Wonderland, asked the Cheshire Cat.

> "That depends a good deal on where you want to get to," said the Cat.
>
> "I don't much care where—" said Alice.
>
> "Then it doesn't matter which way you go," said the Cat.
>
> "—so long as I get *somewhere,*" Alice added as an explanation.
>
> "Oh, you're sure to do that," said the Cat, "if you only walk long enough."

It is "natural" for a species with a well-developed culture to tamper with its biology, and unnatural to refrain. British scientist Dennis Gabor once suggested that our job was to invent the future. And one way or another, as the Cheshire Cat says, we are sure to do that. We may not know who is in the saddle, or exactly where we are going, but we're certainly on our way.

CHAPTER FOURTEEN

THE CULTURAL FUTURE

Ned Ludd Versus The Three Faces of Faust

Civilization is hooped together, brought
Under a rule, under the semblance of peace
By manifold illusion: but man's life is thought,
And he, despite his terror, cannot cease
Ravening through century after century,
Ravening, raging, and uprooting that he may come
Into the desolation of reality.

—W. B. Yeats

In 1779, a simpleton named Ned Ludd lived in the British village of Leicestershire. As was—and still is—the way of boys, young Ned was often teased by the other village children. One day, he chased one of his tormentors into a house, but was unable to catch his quarry; the house contained two weaving machines, however, recently developed for the manufacture of stockings, and Ned Ludd vented his frustration and anger by destroying one of the machines. From then on, whenever a machine broke in Leicestershire, Ludd was blamed. Several decades passed, and by the end of 1811, the war with Napoleon was adding severe economic hardship to the already difficult conditions of newly industrialized Britain. Bands of rioters spread throughout Nottingham and its neighboring districts, destroying machinery in Yorkshire, Lancashire, Derbyshire, and Leicestershire, under the leadership of one "General Ludd," who may well have been mythical.

In any event, the "Luddites" angered the authorities, and struck fear into the newly emerging barons of the Industrial Revolution. The Luddites, too, were driven by anger and fear: anger at the Industrial Revolution itself, which was destroying British cottage industry, raising fear of widespread unemployment as a result of automation, as well as forcing those "lucky" enough to be employed to experience hideous working conditions. The rioting Luddites were harshly suppressed. Following the battle of Waterloo and the defeat of Napoleon, however, Britain suffered a depression, and in 1816, the rioting began again. This time, it spread across all of England, although once more it was eventually put down.

However much sympathy one might feel for the Luddites, we shall gain nothing by mindlessly smashing machinery, neither literal machinery nor the "machinery" of modern society. On the other hand, not all innovations and technologies are good, and not all of our cultural creations deserve respect, or even defense. Some warrant reproach, others (like nuclear weapons) deserve only to be dismantled and perhaps smashed to boot, whereas others (like vaccines, contraception, or literacy) warrant the widest possible dissemination. The quandary of biological Homo sapiens caught in a web of his and her own cultural creation demands a response that is worthy of our designation *sapiens* (the wise) not the inarticulate fury of young Master Ludd.

Unlike Rousseau, we have no reason to extol the noble savage, who probably wasn't all that noble anyhow. And unlike Thoreau, we cannot even retire to Walden Pond (which, incidentally, is now a popular tourist area). We must make the best of our divided existence as cultural animals. Just as we cannot deny our biology, we cannot deny our biological predisposition for culture and the fact that we shall never be able to turn aside from cultural evolution.

But with a clearer vision of what we want for ourselves and

our world, perhaps we can walk more carefully, stubbing our toe here and there, but at least staying on the high ground. Paul Valéry captured the shock, cynicism, and confusion of his generation following World War I when he wrote: "We hope vaguely, but dread precisely; our fears are infinitely clearer than our hopes." In reconciling the hare and the tortoise, one challenge is to fashion hopes that are no less precise, and hence no less attainable, than our fears. With a triumphant gasp of cultural awareness possibly we can even transcend ourselves, assess our position and its causes, and resolve to proceed with a clearer view of ourselves, our needs and our problems. We can evaluate our cultural inventiveness in the light of both our biology and our unique ethical perceptions, and then resolve not to confuse change with progress. After all, it is as absurd to claim that we cannot turn back the clock as it is to harrumph reflexively in opposition to everything that is new. We can, in fact, turn back a clock (even a digital one!) if we elect to do so. Whereas time itself is unstoppable, we can use the time allotted to us as we ourselves see fit. But we cannot turn back on our commitment to cultural solutions to our problems without turning our backs on being human.

Culture is not on trial here, only our uses of it. Given the flexibility of culture, it should actually be more adaptable than our biology alone. As we have seen, the problem is that often, culture is too flexible, serving as a lever that amplifies our biologically given inclinations, hyperextending our capabilities into situations that would be much less momentous if our biology alone were at work. This is why we now push so uninhibitedly on the cultural lever: for thousands of generations, most of today's cultural hyperextenders didn't exist and so our behavior was not harmful. In order for the interface between culture and biology to be benevolent, human behavior—motivated at least in part by biology and acting via culture—must not only *feel* right but also *be* right.

The cultural genie cannot be stuffed back into the bottle, but he can, perhaps, be made to obey. Is it hopelessly utopian and unreasonably idealistic to ask that we learn to measure our culture in terms of human beings rather than measure human beings in terms of machines or institutions? We cannot change "human nature," but we can seek to harmonize our culture with ourselves, always mindful however that what is biological is not necessarily what is good, and knowing also that we may never truly know ourselves.

Each situation must be considered on its own merits in attempting to prescribe remedies. In some cases, notably aggression and overpopulation, our species is necessarily committed to a cultural solution if there is to be any rational solution at all. Of course, all solutions are actually "cultural" in that they must be consciously determined and implemented. The question is whether we should try to move closer to our biology or yet farther away, and the answer must vary according to the tendency in question. Whatever the answers, a clear view of the hare and the tortoise might help us formulate the right questions.

In suggesting possible avenues of reconciliation between our biology and our culture, it is not surprising that being trained as a biologist inclines me to respect the former somewhat more and to criticize the latter. Thus, I would generally have us climb down from our dizzying heights of cultural and technological performance—modulate if not abandon our edifice complex—a recommendation that has been made by others, notably E. F. Schumacher, who has emphasized the value of "intermediate technology," "small is beautiful," and the merits of considering "economics as if people mattered." "Any third-rate engineer or researcher can increase complexity," writes Schumacher. "It takes a certain flair of real insight to make things simple again." In doing so, we might also make ourselves human again.

It is sometimes said among psychotherapists that the patient cannot be brought to a greater level of mental health than that of his or her therapists, just as a guru cannot be expected to lead disciples to more enlightenment than he or she has already attained. Similarly, we can use our diverse cultural innovations for good, but to no more good than we, as human beings, are capable of directing. We design and build societies and machines in the hope of achieving a degree of independence from, and control over, nature. But in the process, we empower ourselves with all the potent leverage of cultural evolution's febrile inventiveness, unleavened by any biologically based restraint. Disciples of Schumacher's approach, and those persuaded, perhaps, by the argument of this book, might argue that for human beings to be psychically sound and materially healthy, and for them to maintain themselves on their planet and with each other on a sustainable basis, they must integrate themselves more comfortably with their cultural evolution. Since we cannot speed up biological evolution, this requires careful scrutiny and often, simplification, of "progress."

On the other hand, many would recommend climbing higher yet. Psychologist B. F. Skinner, for example, while apparently recognizing the awkwardness of our present position, rather predictably takes this alternative view. In his book *Beyond Freedom and Dignity,* he suggested a massive increase in our reliance on cultural artifice, to result ultimately in a "technology of behavior" that will somehow harmonize humanity with its unique self-made environment. Depending on whom one listens to, we therefore might help ourselves by weakening the influence of culture, or by strengthening it. Either way, it seems incumbent on Homo sapiens to change things selectively.

The hare and the tortoise: culture and biology have been racing within us throughout our history. In recent years the hare has been accelerating mightily and the gap between the two has been widening. Reworking the analogy: we have one foot on the tortoise and the other on the hare. As they get farther apart, we are stretched more and more. It's getting awfully uncomfortable.

In *The Future as History,* Robert Heilbroner warned that

> When we estrange ourselves from history we do not enlarge, we diminish ourselves, even as individuals. We subtract from our lives one meaning which they do in fact

> possess, whether we recognize it or not. We cannot help living in history. We can only fail to be aware of it. If we are to meet, endure and transcend the trials and defeats of the future—for trials and defeats there are certain to be—it can only be from a point of view which, seeing the future as part of the sweep of history, enables us to establish our place in that immense procession in which is incorporated whatever hope humankind may have.

Replace "history" with "biology," and you have a reasonable summary of humanity's problem and its promise.

> *We are on Mr. Toad's Wild Ride, and though we may become "velocitized" and insensitive to our speed, or bored, or content to putter about in the apparently motionless comfort of the back seat, at some point we must dare again to look out the window and perceive that we are careening beyond the very human scales of space and time.*
>
> —Walter A. McDougall

According to legend, during the latter part of the sixteenth century there lived in Germany a man named Georg Faust, a magician and conjurer, who blasphemed, bragged about trafficking with the devil, and alternately amazed and irritated his contemporaries. His story was ultimately embellished in theater, opera, and literature, notably by the English playwright Christopher Marlowe, the German poet Johann Wolfgang von Goethe, and the novelist Thomas Mann. In one form or another, the Faust legend has had an extraordinary hold on the Western imagination, and no wonder. Marlowe's *Doctor Faustus* portrays the most widely known version, the tale of a brilliant polymath who turns ultimately to necromancy and thence to a pact with the devil, offering his soul in return for a life of power and voluptuous sensuality. He has his day (actually, twenty-four years) but in the end, pleading and despairing, Faust pays for his excess as he is borne away to eternal damnation by a company of devils.

Mann's *Doctor Faust* is considerably more modern in outlook, not surprising for a novel written in 1947. It tells of Adrian Leverkuhn, an arrogant, sickly musical genius who once again trades his soul, this time in return for creative accomplishment. In Mann's treatment, Leverkuhn's rise clearly parallels the rise of Nazism and the tortured megalomania of overwrought grandiosity and ultimate, inescapable doom.

Both Fausts are cautionary tales, Marlowe's focusing on the excesses of personal ambition and Mann's, on the social sanction of unwarranted power. In each, the central character is a tragic victim in the classic, Aristotelian sense: a hero laid low by a fundamental, inner flaw. And both stories seem to accord with our own flaw as well: Homo sapiens has made a comparable bargain, although the terms were never made explicit nor was the contract ever consciously drawn up. It was, however, sealed with our blood. Faust practiced diabolic hyperextension; ours has been merely cultural. But we have nonetheless been living in the Faust lane.

There is a third face of Faust, however: the one depicted in Goethe's magnificent theater/poem. It presents a different and rather more optimistic vision of Faust, and thus, of humanity. At the play's outset, God and Mephistopheles are debating the merits of Homo sapiens, the latter claiming

> The little god o' the world sticks to the same old way,
> And is as whimsical as on Creation's day.
> Life somewhat better might content him,
> But for the gleam of heavenly light which Thou has lent him:
> He calls it Reason—thence his power's increased,
> To be far beastlier than any beast.

As to the difference between humanity and other animals, Mephistopheles observes

> . . . he to me
> A long-legged grasshopper appears to be,
> That springing flies, and flying springs,
> And in the grass the same old ditty sings.

Would he still lay among the grass he grows in!
Each bit of dung he seeks, to stick his nose in.

A wager is made, with Mephistopheles betting that Faust will succumb to his diabolic temptations, and God betting that ultimately, "I shall lead him to a clearer morning." And of course, this eventually happens. After the furious passions of the Walpurgis Night, a love affair with Helen of Troy, myriad intellectual and scientific accomplishments, political and social success, and great martial triumphs, Faust turns finally to a recognition of human social obligations, whereupon he envisions a community of free people, under his tutelage and protection, living out their lives on free, reclaimed soil. Thereupon, he expresses his satisfaction and his desire that such a moment would last forever. This is what Mephistopheles was waiting for and the terms of his original contract: Faust had not been so much power-mad, as restless, a grasshopper compulsively sticking his nose into "each bit of dung." According to his deal with Mephistopheles, when Faust finally achieved true satisfaction on earth, then and only then would he be damned. Yet, when that finally occurs and Faust dies, God intervenes and has him transported to heaven as reward for his ultimate recognition of the highest goal of human existence, responsibility for others.

The limits of human potential are glimpsed, if at all, with much greater difficulty than are the outlines of our folly. In seeking to become angels, warned Pascal, we risk becoming less than men. His warning can be inverted as well: in seeking to become angels, we cannot become more than men or women. Perhaps Mephistopheles and a cynical reading of human biological evolution are correct, and we are doomed eternally to sing "the same old ditty," big-brained grasshoppers without even the good grace to recognize and act within our nature. On the other hand, perhaps we can reach for a deeper humanity, thumb our noses at Mephistopheles and at the grasshopper within us, saying "no" to the siren whisperings from our genes—or alternatively, to certian blandishments of our culture—whenever this seems necessary to achieve a truly sapient *Homo*.

Any effort to reveal human nature must unavoidably grapple with and acknowledge fundamental uncertainties. At minimum, we seem doomed to a kind of existential agnosticism, never knowing—and perhaps, never able to know—ourselves. What, then, should we do? It would be foolhardy if not irresponsible to let "nature" take its course, either our own biological nature, or the nature of human culture as currently constituted, because nature isn't necessarily good, and moreover, nature at this point has no course independent of the one we plot. We are also deeply ignorant both of our limitations and our capabilities. For a useful recommendation, we might turn from Faust to another German, Hans Vaihinger, who developed the philosophy of *"Als Ob"*: As If. Vaihinger emphasized that we make the world meaningful—indeed, tolerable—by living with certain essential fictions. We behave As If we are not personally going to die; we behave As If the sensory impressions that we receive are accurate reflections of an underlying reality; we behave As If a just and benevolent god will reward virtue. (Or for some of us, As If it carries its own reward, and for others As If such concerns can be ignored altogether.) We might also want to consider behaving As If we are free to say "no" or "yes" to our biology and our culture, as we choose, depending on whatever values we embrace. Wishful thinking cannot hold off personal death, or manipulate the insensate world, or create god from indifferent chaos. But when applied to human action on our human-dominated planet, behaving As If can become a self-fulfilling prophecy, something that empowers us in proportion as we believe it.

It remains to be seen, of course, whether Homo sapiens really does want to seize the future, and to do so in a uniquely human way, not just as the agents of natural selection acting around us and through us. "One of the most striking things about the struggle for freedom from intentional control is how often it has been lacking," wrote B. F. Skinner, in his *Beyond Freedom and Dignity*.

"Many people have submitted to the most odious religious, governmental, and economic controls for centuries." The theme recurs in some of our most renowned works of fiction as well. In "The Grand Inquisitor" chapter of Dostoevsky's *The Brothers Karamazov,* for example, Christ returns to earth during the Spanish Inquisition and is informed that "nothing has ever been more insupportable for a man and a human society than freedom." And in describing the effects of the Inquisition, we are told that "men rejoiced that they were again led like sheep, and that the terrible gift that had brought them such suffering was, at last, lifted from their hearts." At least, according to the Inquisitor.

On the other hand, that "terrible gift," human freedom, has motivated some of the most courageous and persistent actions of all history, from political revolution to scientific discovery.

Our challenge is to use our unique freedom as human beings to humanize culture, technology, science, and society; in doing so, our reward is that we shall have humanized ourselves. The struggle to maintain selfhood in the face of such dangers is awesome, and important, but it is not altogether new. After all, the effort to avoid being dehumanized in the twentieth century is not all that different from the earlier effort to avoid being "de-humanized" by a saber-tooth. We won that battle; we can also win this one.

There are, of course, some important differences between the threat of the saber-tooth and the threat of technology run wild: whereas de-humanization by the saber-tooth meant death of the body, dehumanization by our own creations means death of the soul. The danger in the former case was physical, immediate, and easily recognized, not different from threats that our primitive ancestors faced—successfully—for hundreds of millions of years. Appropriate responses were therefore easily evoked. The danger in the latter case is no less real, but nonetheless diffuse and arguable, often clothed in alluring Faustian promises of greater material abundance, personal accomplishment, or physical power. As with the problem of supernormal releasers, whose dangers to us are amplified, paradoxically, by their diffuseness, the threat of modern-day dehumanization is enhanced by its deceptive allure and subtle nuance.

Goethe's version is widely acknowledged to be the greatest Faust of all. But it is up to us whether Faust will be our text, and if so, which Faust. In this book, we have sought to trace the history and terms of humanity's bargain with evolution, both biological and cultural. We began with an image of murder, many eons ago. How appropriate, then, that we should end with the possibility of redemption.

Notes and References

One of the nice things about reading (and writing) a "trade" book as opposed to an academic tome is that you don't have to get bogged down by excessive references. So, when a poem or book's title was mentioned in the text, I have not repeated it here. On the other hand, one pleasure of reading any book is to use it as a jumping-off place for more information, just as it is another pleasure to recommend favorite sources to anyone who will listen. So, in this spirit, I offer the following notes and references: not in compliance with a scholarly ritual, but as one might introduce old friends to new ones, in the hope that they will enjoy each other's company.

CHAPTER TWO

More information regarding Tycho Brahe can be found in John Gade's *The Life and Times of Tycho Brahe* (1947, Princeton University Press: Princeton, New Jersey). There have been an enormous number of books written about evolution, many of them quite good, and some truly excellent. For scholarly textbook–type introductions—but nonetheless readable ones—my own favorites are *Evolution* (1977, W. H. Freeman: San Francisco) by the great Russian-born geneticist Theodosius Dobzhansky, *The Theory of Evolution* (1966, Penguin: Harmondsworth, England) by the British mathematical ecologist John Maynard Smith, and *Processes of Organic Evolution* (1977, Prentice-Hall: Englewood Cliffs, New Jersey) by the American botanist G. Ledyard Stebbins. Stebbins's *Darwin to DNA: Molecules to Humanity* (1982, W. H. Freeman: San Francisco) is more recent and has more material. These books emphasize the process of evolution, notably natural selection. For more focus on the history of the concept, try Loren Eiseley's *Darwin's Century* (1958, Doubleday: Garden City, New York). Peter Bowler's *Evolution: The History of an Idea* (1984, University of California Press: Berkeley) is—not surprisingly—less readable than Eiseley, but more scholarly. Two beautifully written classics that treat evolution in relation to human beings are Eiseley's *The Immense Journey* (1957, Random House: New York) and Garrett Hardin's *Nature and Man's Fate* (1959, Rinehart: New York). For a solid introduction to the actual history of evolution itself—notably, the process of evolutionary change among our distant ancestors—try Edwin H. Colbert's justly celebrated *Evolution of the Vertebrates* (1980, John Wiley: New York).

The literature on human evolution, and on primates, is also enormous. For starters, you might want to try Elwyn L. Simons's *Primate Evolution: An Introduction to Man's Place in Nature* (1972, Macmillan: New York) and also Alison Jolly's *The Evolution of Primate Behavior* (1972, Macmillan: New York). There has been much speculation but—not surprisingly—few facts discovered regarding the actual origin of life on earth; what is known is ably reviewed in *The Nature and Origin of the Biological World* (1982, Halsted: New York) by E. J. Ambrose. My favorite treatment of human history has always been H. G. Wells's *The Outline of History,* which, as the subtitle indicates, is no less than a "plain history of life and mankind" (1920, Macmillan: New York). It's also a treat to look through Hilaire Belloc's companion to Wells (1926, Sheed & Ward: London). Belloc, incidentally, summarized the West's reliance on culture—or rather, technology—when he wrote the following trenchant lines at the time India was seeking to overthrow British rule:

> Whatever happens, we have got
> The Maxim gun, and they have not.

CHAPTER THREE

For more on Lamarck, try *The Spirit of System: Lamarck and Evolutionary Biology* (1977, Harvard University Press: Cambridge, Mass.) by Richard W. Burkhardt. Paul MacLean's evocation of the three-part human brain occurs in his "A Triune Concept of the Brain and Behavior," part of a congress held at Queen's University in Kingston, Ontario, and published by the University of Toronto Press in 1973. Alvin Toffler's *Future Shock* (1970, Random House: New York) contains some themes similar to the present book, although he does not identify the culture/biology conflict as the principle cause, and he is much more enthusiastic about the human-made future than I am. For more on the history and consequences of our mobility, I recommend *Human Migrations: Patterns and Policies* (1978, Indiana University Press: Bloomington) edited by William McNeill and Ruth S. Adams.

For basic material on sociobiology, I immodestly recommend my own textbook *Sociobiology and Behavior* (1982, Elsevier: New York), and my trade book *The Whisperings Within* (1979, Harper & Row: New York). Edward O. Wilson's massive, now-classic compilation was titled *Sociobiology: The New Synthesis* (1975, Harvard University Press: Cambridge, Mass.) and was followed by his *On Human Nature* (1978, Harvard University Press: Cambridge, Mass.) As with most efforts to "biologize" human behavior, however, sociobiology has also been controversial; for two critiques—which to my mind unfairly lump sociobiology with racist eugenics and misguided social Darwinism—see Richard C. Lewontin, S. Rose, and L. J. Kamin's *Not in Our Genes* (1984, Pantheon: New York) and Stephen Jay Gould's *The Mismeasure of Man* (1981, W. W. Norton: New York). An intellectually challenging, mathematical argument for the connection between human evolution and culture is given by Charles Lumsden and Edward O. Wilson in their *Genes, Mind and Culture: The Coevolutionary Process* (1981, Harvard University Press: Cambridge, Mass.). A more readable treatment, for nonspecialists, is *Promethean Fire: Reflections on the Origin of the Mind* (1983, Harvard University Press: Cambridge, Mass.) by the same formidable pair. Robert Boyd and Peter J. Richerson have recently presented a series of models of *Culture and the Evolutionary Process* (1985, University of Chicago Press: Chicago), in which they analyze those situations in which natural selection would favor different capacities for cultural transmission. Finally, the Weston La Barre quote is from his magisterial *The Ghost Dance: Origin of Religion* (1970, Doubleday: Garden City, New York). For what it is worth, La Barre is my favorite anthropologist, a brilliant and iconoclastic scholar whose work deserves more attention than it has received.

CHAPTER FOUR

For a traditional ethological approach to the subject, see Margaret Bastock's *Courtship* (1967, Aldine: Chicago) and the relevant chapters of Irenaus Eibl-Eibesfeldt's well-illustrated text, *Ethology: The Biology of Behavior* (1975, Holt, Rinehart & Winston: New York). Sociobiologic views of sexual behavior can be found in my own *Sociobiology and Behavior* (1982, Elsevier: New York), and in Martin Daly and M. Wilson's *Sex, Evolution and Behavior* (1978, Duxbury: North Scituate, Mass.). A recent volume of scientific papers, edited by Cambridge University's Patrick Bateson, is also available: *Mate Choice* (1983, Cambridge University Press: New York). For the biology of maleness, see Robert L. Smith, ed., *Sperm Competition and the Evolution of Animal Mating Systems* (1984, Academic Press: Orlando, Fla.). Material on orchid/wasp mimicry, as well as other similar systems, is presented in Wolfgang Wickler's beautifully illustrated *Mimicry in Plants and Animals* (1968, McGraw-Hill: New York).

Regarding a sociobiologic view of human pair bonding, see my own *The Whisperings Within* (1979, Harper & Row: New York), and *The Evolution of Human Sexuality* (1979, Oxford University Press: New York) by anthropologist Donald Symons. Compare these treatments with reports by nonbiological writers, such as *A History of Courting* (1955, E. P. Dutton: New York) by Ernest Sackville Turner, or the more recent *Hands and Hearts: A History of Courtship* (1984, Basic: New York) by Ellen K. Rothman. German ethologist Irenaus Eibl-Eibesfeldt provides a traditional ethological perspective on human pair bonding in his *Love and Hate* (1972, Holt, Rinehart & Winston: New York). Probably the most influential—and popular—attempt to biologize human behavior, including a good exposition of our penchant for face-to-face lovemaking, was *The Naked Ape* (1965, McGraw-Hill: New York) by ethologist Desmond Morris. For inbreeding depression, see the massive (965-page!) *Genetics of Human Populations* (1971, W. H. Freeman: San Francisco) by Luigi L. Cavalli-Sforza and William F. Bodmer. For a sociobiologic perspective on incest: the best source is Joseph Shepher's *Incest: A Biosocial View* (1983, Academic Press: New York). And Randy and Nancy Thornhill present their research in "Human Rape: An Evolutionary Perspective," which appeared in the journal *Ethology and Sociobiology* (1983, 7: 137–173).

CHAPTER FIVE

Emma Goldman's "Marriage and Love" appears in her *Anarchism and Other Essays* (1911, Mother Earth Publishing Co.: New York). Friedrich Engels, collaborator with Karl Marx, wrote a treatment of the human family and the origin of female oppression that is remarkably "sociobiologic" in orientation: *The Origin of the Family, Private*

Property and the State (1972, International Publishers: New York). For the sociobiology of male-female differences, see my *Sociobiology and Behavior* (1982, Elsevier: New York) or Edward O. Wilson's *Sociobiology: The New Synthesis* (1975, Harvard University Press: Cambridge, Mass.). Anyone interested in feminism should consult the following if he/she hasn't already: *The Second Sex* (1952, Alfred A. Knopf: New York) by Simone de Beauvoir; *The Feminine Mystique* (1963, W. W. Norton: New York) by Betty Friedan; *The Female Eunuch* (1971, McGraw-Hill: New York) by Germaine Greer; and *Sexual Politics* (1970, Doubleday: Garden City, New York) by Kate Millett. It is noteworthy that Greer has recently paid homage to biology in her essay on the struggles between feminism and reproduction, *Sex and Destiny: The Politics of Human Fertility* (1984, Harper & Row: New York). Sociobiologist—and feminist—Sarah Blaffer Hrdy (not a misprint, no "a") provides a refreshing perspective on male-female evolution in her *The Woman That Never Evolved* (1981, Harvard University Press: Cambridge, Mass.).

Before Gilligan, the standard and rather male-centered view of human moral development was that of Lawrence Kohlberg, ably and influentially expressed in his *The Philosophy of Moral Development: Moral Stages and the Idea of Justice* (1981, Harper & Row: New York). In her *The Hearts of Men: American Dreams and the Flight from Commitment* (1983, Doubleday: Garden City, New York), Barbara Ehrenreich, like Gilligan, makes an argument that could have been designed by an evolutionary biologist, but wasn't.

CHAPTER SIX

Richard Alexander has written very effectively on the evolutionary aspects of human social organization, in his *Darwinism and Human Affairs* (1979, University of Washington Press: Seattle). For James Lloyd's work on fireflies, see his article "Aggressive Mimicry in *Photuris* Firefly *Femmes Fatales*," which appeared in the journal *Science* (1965, 149: 653–654). A fine review of the phenomenon of imprinting is available in W. Sluckin's *Early Learning in Man and Animals* (1970, Allen & Unwin: London). The experiment regarding musical preferences of rats is recounted in a chapter by social psychologist Robert Zajonc (rhymes with "science") titled "Attraction, Affiliation and Attachment," in *Man and Beast: Comparative Social Behavior*, edited by John F. Eisenberg and Wilton S. Dillon (1971, Smithsonian Institution Press: Washington, D.C.). The double standard among swallows is described in an article by Mike and Inger Beecher, "Sociobiology of Bank Swallows: Reproductive Strategy of the Male," in the journal *Animal Behaviour* (1979, 205: 1282–1285).

Excellent research on animal alarm calling has been reported by Paul W. Sherman, "Nepotism and the Evolution of Alarm Calls," in *Science* (1977, 197: 1246–1253). I also heartily recommend Donald T. Campbell's article "On the Conflicts between Biological and Social Evolution and between Psychology and Moral Tradition," appearing in the journal *American Psychologist* (1975, 30: 1103–1126). Sociologist Pierre van den Berghe treats human group identification from an evolutionary perspective in his prize-winning *The Ethnic Phenomenon* (1981, Elsevier: New York). The primitive biology underpinning modern-day nationalism is explored in *The Caveman and the Bomb* (1985, McGraw-Hill: New York) by David P. Barash and Judith Eve Lipton. And for an introduction to the very serious game of Prisoner's Dilemma—as well as a way out—see Robert Axelrod's *The Evolution of Cooperation* (1984, Basic: New York).

CHAPTER SEVEN

Frank Beach's warning to comparative psychologists, with the intriguing title "The Snark Was a Boojum," appeared in the journal *American Psychologist* (1950, 5: 115–124). The founder of modern ethology, Konrad Z. Lorenz, tells some marvelous animal stories—many of them related to aggression—in his justly famous *King Solomon's Ring* (1952, T. Y. Crowell: New York). Lorenz's *On Aggression* (1966, Harcourt, Brace and World: New York), although a bit dated in its "good of the species" arguments, nonetheless does a fine job of presenting the ethological view of animal and human nastiness. In part to counter such notions, the renowned American anthropologist Ashley Montagu edited *Man and Aggression* (1973, Oxford University Press: New York), which argued, aggressively, that we aren't instinctively aggressive at all. In his book *Aggression* (1958, University of Chicago Press: Chicago), animal psychologist John Paul Scott brought together data on the role of environmental factors—especially social disruption—in producing aggression among animals. Albert Bandura's *Aggression* (1973, Prentice-Hall: Englewood Cliffs, New Jersey) is an acclaimed social learning approach, and in *Frustration and Aggression* (1961, Yale University Press: New Haven, Conn.), psychiatrist John Dollard and others develop the thesis that frustration leads to aggressive behavior. For some biological views of aggression,

Robert Ardrey's books (*African Genesis,* 1961; *The Territorial Imperative,* 1966; and *The Social Contract,* 1970, all published by Atheneum: New York) have been criticized as glib and a bit naive; they are certainly entertaining, however, and worth reading nonetheless. For a somewhat more scholarly biological view of human war, representing the ethological perspective, see I. Eibl-Eibesfeldt's *The Biology of Peace and War* (1979, Viking: New York). And for a sociobiological flavor, try William Durham's article "Resource Competition and Human Aggression, Part I: A Review of Primitive War," in the journal *Quarterly Review of Biology* (1976, 51: 385–415).

A marvelous, accessible treatment of human spacing behavior was presented by anthropologist Edward T. Hall in *The Hidden Dimension* (1969, Doubleday: Garden City, New York). Microbiologist-humanist Rene J. Dubos has written about the range of human adaptability in *Man Adapting* (1965, Yale University Press: New Haven, Conn.) and *Of Human Diversity* (1974, Crown: New York). Johan Huizinga's *Homo ludens: A Study of the Play Element in Culture,* was published in 1950 by Beacon Press (Boston).

CHAPTER EIGHT

The imbalance between human biology and human culture is most pronounced—and most perilous—in the arena of nuclear weapons. For an elaboration of the hare and tortoise applied to nuclear war and the arms race, see *The Caveman and the Bomb: Human Nature, Evolution and Nuclear War* (1985, McGraw-Hill: New York) by David P. Barash and Judith Eve Lipton. A notable treatment of the arms race from a social psychological perspective was written by psychiatrist-psychologist Jerome Frank in 1965, then reprinted in 1982 as *Sanity and Survival in the Nuclear Age* (Random House: New York). Psychologist–public official Ralph White's *Fearful Warriors* (1984, Free Press: New York) ably describes the impact of fear and pride in international affairs, highlighting the psychological dimension in superpower relations. For another view of the psychological impact of nuclear weapons on human thought, see Joel Kovel's *Against the State of Nuclear Terror* (1984, South End Press: Boston). The Rebecca West quotation is from her *Black Lamb and Grey Falcon* (1940, Viking Penguin: New York), and the research by William Beardsley and John Mack is discussed in their chapter "The Impact on Children and Adolescents of Nuclear Developments," in *Psychological Aspects of Nuclear Developments,* Task Force Report #20, American Psychiatric Association, Washington, D.C. A powerful argument against the technological, strategic, and economic feasibility of Star Wars can be found in *The Fallacy of Star Wars* (1984, Vintage: New York) by physicists Richard Garwin, K. Gottfried, and H. Kendall. Perhaps the best treatment of "nuclear despair" is by Joanna Rogers Macy in *Despair and Personal Power in the Nuclear Age* (1983, New Society: Philadelphia).

CHAPTER NINE

Probably the most influential book on the threat and reality of overpopulation was Paul Ehrlich's nifty little *The Population Bomb* (1968, Ballantine: New York). For more meat, try Paul Ehrlich and Anne Ehrlich's *Population, Resources, Environment* (W. H. Freeman: San Francisco). Among the many books written about this issue, I particularly recommend Lester R. Brown's *In the Human Interest* (1974, W. W. Norton: New York) and *Ecocide and Population* (1971, St. Martin's: New York), edited by M. E. Adelstein and J. G. Pivall. The other side—an optimistic view that I consider misleading and dangerous—is represented by Julian L. Simon and Herman Kahn, eds., *The Resourceful Earth* (1984, Basil Blackwell: New York). The sad story of the Kaibab deer can be found in John P. Russo's *The Kaibab North Deer Herd,* a 1964 publication of the State of Arizona Game and Fish Department, Phoenix. For more on the demographic transition among human beings, see *Population Pressure and Cultural Adjustment* (1979, Human Sciences Press: New York) by Virginia Abernathy. And perhaps the best source for examples of density-dependence in animals is V. C. Wynne Edwards's *Animal Dispersion in Relation to Social Behavior* (1962, Hafner: New York). Professor Wynne Edwards attributed these effects to selection acting at the level of groups—not a popular viewpoint these days—but his work is a rich source of references and data, regardless of interpretation.

John Calhoun's work on Norway rats is recounted in his *The Ecology and Sociology of the Norway Rat* (1963, U.S. Public Health Service Publication #1008, Bethesda, Maryland). John Christian and David E. Davis review the relationship of adrenal size to stress and population density in their article "Endocrines, Behavior, and Population," which was published in the journal *Science* (1964, 146: 1550–1560). H. M. Bruce discusses "the Bruce effect" in "Smell as an Exteroceptive Factor," in the *Journal of Animal Science* (1966, supplement #25: 83–89). And for the classic account of lemming behavior and ecology, without the Walt Disney myths, try *Voles, Mice and Lemmings* (1942, Clarendon Press: Oxford) by the noted ecologist Charles S. Elton.

CHAPTER TEN

Two very important papers on the relationship of Homo sapiens to its environment are Lynn White's "The Historical Roots of Our Ecological Crisis" (reprinted in *Ecocide and Population,* referenced in Chapter 9) and Garrett Hardin's "The Tragedy of the Commons," in the journal *Science* (1961, 162: 1243–1248). The quote from Susanne Langer appears in her *Philosophy in a New Key* (1951, Harvard University Press: Cambridge, Mass.). When it comes to the possibility of establishing a healthy practical rapport between people and their economic needs, perhaps the most insightful thinker has been Ernst Friedrich Schumacher, especially his *Small Is Beautiful: Economics as if People Mattered* (1973, Harper & Row: New York) and *Good Work* (1979, Harper & Row: New York). The C. D. Darlington quote is from his *The Evolution of Man and Society* (1969, Simon & Schuster: New York). Readers concerned about the extinction of animal and plant species might want to consult Paul and Anne Ehrlich's *Extinction: The Causes and Consequences of the Disappearance of Species* (1981, Random House: New York). And for a chilling review of the prospects for nuclear winter, the most devastating environmental disaster of all, read *The Cold and the Dark* (1983, W. W. Norton: New York), edited by Paul Ehrlich, Carl Sagan, Donald Kennedy, and Walter Orr Roberts.

CHAPTER ELEVEN

Noted biologist John Tyler Bonner, an expert on the embryology and behavior of slime molds, has also written *The Evolution of Culture in Animals* (1980, Princeton University Press: Princeton, New Jersey). Pierre Teilhard de Chardin's enthusiasm for technology is expressed in his *The Phenomenon of Man* (1959) and *The Future of Man* (1964), both published by Harper (New York). Readers wanting more of Alexander Herzen might try consulting his sometimes provocative *From the Other Shore* (1956, George Braziller: New York). Writing in that wonderful leftist stalwart *The Nation,* on April 27, 1985, Robert Engler provided a very useful and disturbing review of human technology, "Technology Out of Control." In his important book *Normal*

Accidents (1984, Basic: New York), C. Perrow developed the thesis that overly complex technology leads inevitably to breakdowns and errors, some of them potentially catastrophic. Similar sentiments, but more gloomy and abstract, oriented more around a philosophical and theological perspective, can be found in the varied writings of France's Jacques Ellul, notably his *The Technological Society* (1964, Alfred A. Knopf: New York) and *The Betrayal of the West* (1978, Seabury: New York). Prolific and versatile historian William McNeill recently wrote what is probably the best account of the history of technology and armed force, *The Pursuit of Power* (1983, Basil Blackwell: New York). For a fine, level-headed, and well-balanced treatment of the role of technology in recent human history, see Victor Ferkiss, *Technological Man* (1969, George Braziller: New York), but if you want startling brilliance and erudition combined, the master is Lewis Mumford, who has specialized in architecture, written a critical study of Herman Melville, and produced several classics on technology and human society, including *Technics and Civilization* (1963) and *The Myth of the Machine* (1938), both published by Harcourt, Brace and World (New York).

CHAPTER TWELVE

To my mind, the most effective introduction to existentialism, the philosophy of alienation, is still William Barrett's *Irrational Man* (1958, Doubleday: Garden City, New York). Martin Buber's *I and Thou* (1970, Charles Scribner's Sons: New York) should not be missed, nor should Ian McHarg's *Design with Nature* (1971, Doubleday: Garden City, New York). Biologist W. D. Hamilton's article "Geometry for the Selfish Herd" appeared in the *Journal of Theoretical Biology* (1971, 31: 295–311). For a study of alienation as a major component of Karl Marx's analysis of modern society, see *Alienation, Praxis and Techne in the Thought of Karl Marx* (1976, University of Texas Press: Austin) by K. Axelos. Alienation is also very much on the mind of many specialists, from many different fields, as they survey the human condition. For philosophical perspectives, see I. Feuerlicht's *Alienation: From the Past to the Future* (1978, Greenwood Press: Westport, Conn.) and Morton Kaplan's *Alienation and Identification* (1976, Free Press: New York). Brian Baxter has reviewed the issue of alienation at the workplace in his *Alienation and Authenticity* (1982, Tavistock: New York). And for the impact of alienation on adolescents and the young in American Society, see Kenneth Kenniston's *The Uncommitted* (1965, Harcourt, Brace and World: New York).

CHAPTER THIRTEEN

The Emerson epigraph is from his "Ode, inscribed to W. H. Channing," which can be found in *The Selected Writings of Ralph Waldo Emerson* (1950, Modern Library: New York), edited by Brooks Atkinson. I have reviewed the biology (as well as other aspects) of aging in my *Aging: An Exploration* (1983, University of Washington Press: Seattle). Anthropologist Ruth Benedict, renowned for her *Patterns of Culture,* turned her attention to the problem of the human races in collaboration with Gene Weltfish in *Race: Science and Politics* (1959, Viking: New York). For sickle-cell anemia, try A. Cerami and E. Washington, *Sickle Cell Anemia* (1977, The Third Press: New York). The quote from economist-historian Robert L. Heilbroner is from his *The Future as History* (1960, Harper: New York).

CHAPTER FOURTEEN

For more on the Luddites, try either Frank O. Darvall's *Popular Disturbances and Public Order in Regency England* (1934, Oxford University Press: Oxford) or Frank Peel's *The Risings of the Luddites, Chartists and Plug-drawers* (1968, Cass: London). The latter has the advantage of an introduction by the brilliant and acerbic British social historian E. P. Thompson, who is also founder of the END (European Nuclear Disarmament) campaign. Finally, for a gentle yet stirring evocation of the human potential, try Rene J. Dubos's *Beast or Angel?* (1974, Charles Scribner's Sons: New York).

Index

Acknowledgments

Writers of nonfiction books often have a long list of people to thank, frequently after pointing out that books are rarely written by the author alone. Well, for better or worse, this one pretty much was. I even typed it myself. On the other hand, Drs. Barbara and Morris Lipton made helpful comments, and Dan Frank, my editor at Viking, and his assistant, André Bernard, prodded me beneficently to clarify arguments and prune unneeded verbiage. Nanelle Rose Barash (whose first four months overlapped the last four of this book) did much to distract me at critical junctures, thereby doubtless keeping the flow of ideas from becoming excessive.